The SECRET LANGUAGE of the BODY

Regulate your nervous system, heal your body, free your mind

JENNIFER MANN & KARDEN RABIN

HARPERONE
An Imprint of HarperCollins*Publishers*

The content of this book is intended to inform, entertain, and provoke your thinking. This is not intended as medical advice. Neither the author nor the publisher can be held responsible or liable for any loss or claim arising from the use, or misuse, of the content of this book. Names and identifying details of patients have been changed to protect their identity.

The QR codes in this book contain links to websites hosted by third parties. HarperCollins Publishers cannot guarantee that they will remain functional following publication of this edition.

HarperCollins books may be purchased for educational, business, or sales promotional use. For information, please email the Special Markets Department at SPsales@harpercollins.com.

Originally published as *The Secret Language of the Body* in Ireland in 2024 by Thorsons.

First HarperOne paperback published 2024

Design adapted from the Thorsons edition
Practice illustrations by Julian Gower
Diagrams by Liane Payne

Library of Congress Cataloging-in-Publication Data has been applied for.

ISBN 978-0-06-338238-1

ISBN 978-0-00-866765-8 (international edition)

24 25 26 27 28 LBC 5 4 3 2 1

To Gillian, my little women, and Samantha
- K.R.

To my love Yian, my precious boy Leo and to
my nervous system that taught me that I
am resilient, not broken
- J.M.

Contents

Introduction 1
Reader's Guide 17

Part I – Mind

Chapter 1: Awareness - Listening 29
Chapter 2: Interruption - Switching 57
Chapter 3: Redesign - Distancing 91

Part II – Body

Chapter 4: Awareness - Translating 119
Chapter 5: Interruption - Modifying 159
Chapter 6: Redesign - Settling 205

Part III – Human

Chapter 7: Awareness - Attuning 233
Chapter 8: Interruption - Tending 267
Chapter 9: Redesign - Bonding 291

Conclusion 323
Notes 335

Introduction

If you're stuck, we can help

You are not broken.

Jen and Karden

For change to happen in life, you need either inspiration or desperation. While we sincerely hope that you are feeling inspiration it is likely many of you are being driven by desperation, like us and many others who come to this work. You may be suffering from stress, anxiety and unresolved trauma. You may also feel hopeless on how to heal, how to feel safe in your own body, how to be at peace in your mind, and just eager to be living your life instead of simply surviving it. If you've picked up this book, beloved reader, you have undoubtedly set foot on a path of healing. And, as you move through these coming pages at your own pace, we hope that you will discover the wisdom that already lies within you and arrive at a place that feels less like struggle and more like home.

Remember this: there is a boundless healing capacity inside you, an inherent strength, awareness and intuition that is more powerful than you can imagine. This book is your invitation to awaken the vast potential hidden within the language of your own body, become fluent in your body's inner dialogue and *finally* heal.

You may be reading this book because you are looking for actionable, practical and effective tools for healing yourself from chronic stress, anxiety, workaholism, depression, addiction, disordered eating, procrastination, physical pain or a host of stress-based illnesses like irritable bowel syndrome (IBS), allergies, asthma, skin conditions, chronic fatigue syndrome, migraines, post-viral syndromes and many more. Something inside you recognises that you can heal yourself. By the time you are picking up this book, it's likely you've already tried conventional doctors, exercise, diet, functional medicine, supplements, yoga, cannabidiol (CBD) oils, deep tissue therapy, shiatsu, reiki, reflexology, acupuncture, cupping, hot tubs, saunas, aromatherapy, gong baths, salt scrubs, singing bowls, crystals, autonomous sensory meridian response (ASMR), binaural beats, chamomile tea, cognitive behavioural therapy (CBT), ayahuasca, talk therapy and a hundred and one different ways to breathe. Although some of these approaches may have helped, something is still incomplete, and you still don't feel the way you want to feel or live the way you want to live. The person you see in the mirror does not feel quite like you. We intimately know where you are because we were there too. Stressed, sick and stuck - often feeling hopeless - but with a persistent, inner knowing that there *must* be a way out.

Spoiler alert: We can help! Through our personal healing journeys and the many teachers we have learned from, we found the way out and by using the exercises in this book have already helped thousands of other people to recover as well. Through this book, we can help you to heal too. Chances are, if you've been through the many treatment approaches above, you may have considered that your nervous system may be at the root of your problems. But your nervous system isn't just an inert wiring system in your body that likes massages and chamomile tea. It's an awe-inspiring, sophisticated network that interprets, processes and remembers impressive amounts of detailed information at all times. It coordinates every function of your existence including, but not limited to, helping you breathe and making your heart beat, manufacturing your experience of consciousness, and helping you decide whether you want to buy sunflowers or peonies for your living room.

We are about to take you on a journey of learning how to listen to and 'speak' the somatic non-verbal, ancient, wild, beautiful and sometimes scary language of your nervous system. If you can't hear that language, much less 'speak' it, you cannot hope to influence it, which is the fundamental missing piece needed to repair your nervous system and heal your health and life. To put this into perspective, imagine you are travelling through a foreign country and want to ask someone how an ancient monument was built. How would you do this if you don't speak the language? It would be challenging, wouldn't it? Now imagine that this country is your own body, and instead of wanting to learn how an ancient monument was built, you want to understand the cause of your anxiety. How can you do this if you don't speak the language your anxiety is communicating to you? Despite the helpful insights you and your therapist may have unearthed or your scientific understanding of the hypothalamic-pituitary-adrenal (HPA) axis, the truth is you can't fully understand the cause of your anxiety if you don't speak the language of the nervous system.

It's essential to understand the basic logic and purpose of your nervous system from an evolutionary and functional viewpoint to truly grasp why it has so much power over you (and why - currently - you have so little power over it). Consider the importance of logic and purpose in musical notation. The words and symbols only make sense when you learn that they were designed to express sounds to create music. Without that context, even though you can see the notations, they are meaningless. Similarly, understanding the nervous system's logic and purpose helps to make learning the language much more straightforward.

This book is devoted to teaching you, in the most practical of ways, how to regulate your nervous system and heal yourself. The method you will learn is called A I R, which stands for awareness, interruption and redesign. The goal of A I R is to develop the internal **awareness** needed to detect the language of your nervous system within your mind, your body and your human aspects; to **interrupt** the unhelpful and cyclical patterns that are keeping you stuck; and to **redesign** those patterns by fluently speaking the secret language of the body. By applying A I R and engaging in the practices at the end of each chapter, you will gradually learn to tap into the power of your nervous system for self-healing. In the same way that learning a new language allows you to navigate a foreign country and unlock the secrets and richness of that country's people and culture, you will be able to do that in the native soil of your own home: your mind, your body and your human. As you move through the chapters you will find that the exercises build on each other. Feel free to make your own way through these, take notes on what works for you and tailor them to your path. Before you can learn this new language, though, we want to help you understand this 'foreign country' of your nervous system, why it reacts the way it does, and how speaking this language and working with A I R actually works.

A I R	**Awareness →**	**Interruption →**	**Redesign**
Mind	listening	switching	distancing
Body	translating	modifying	settling
Human	attuning	tending	bonding

So, let's begin.

Your nervous system exists to help your body produce physiological adaptations to ensure your survival. In other words, you have a nervous system so you can have your best chance at staying alive. In the binary programming language of your nervous system the persistent question is: am I *safe* or *unsafe*? And every single thing it does, from determining your hormone ratios to choosing a romantic partner, is *always* guided by this all-important binary question and the priority of keeping you safe and alive. If it perceives what's happening to you as a threat (even if it's not), it will deploy a predictable set of actions along a spectrum of survival (stress) responses biologically hardwired into you through millions of years of evolution. We know not all stress is bad, and after millions of years you would think this biological programme of the body would be fine-tuned beyond safe/unsafe but, in reality, there are complexities that can trip it up.

The first complexity is that your nervous system is not appropriately equipped for modern-day stressors. You are designed to survive the African savanna and lions, not the urban jungle and the internet. Your nervous system *literally* deploys the same survival responses to help you pay your overdue credit-card bill as it did to escape an angry rhinoceros 25,000 years ago. It's a bewildering mismatch. A response that would be carried out and metabolised by running for your life now gets repressed and stagnant in your body as you sit in your office chair paying bills and answering emails. Worse yet, the mismatch is compounded by the relentless barrage of modern-day stressors. Where our ancestors were only confronted with threats a few times a day or week, we are confronted by threats every time our ever-present phone sends us a push notification about a message, an email or the most recent shocking news event. Essentially, modern humans are using Stone Age nervous system technology to navigate a world that moves at the speed of light, inevitably leading to compounded stress in the body.

The second complexity is that your survival functions are directed by your autonomic nervous system (ANS) which, until you learn the language of the nervous system, is usually unconscious, automatic and out of your cognitive control. That's why – as you have probably noticed much to your displeasure – whenever you get triggered, even if your conscious mind wants you to remain calm, your ANS will ignore it and repeat the negative pattern it's accustomed to deploying. So, without access to the language of your nervous system, using logic to tell yourself to calm down and expecting success is like expecting someone who doesn't speak your language to understand what you're saying!

> **Calm is felt in the body, not in the mind.**

The third complexity is that your nervous system's ability to serve its purpose of protecting you is only as good as its programming. This programming is built on your past experiences. While most of us can barely remember much before our fifth birthday, our bodies have been accumulating and storing experiences and feelings for much longer. In fact, as early as week six in utero your nervous system starts developing and gathering information to keep you alive.[1] The cognitive function part of the brain is actually the last to form in a developing baby. While they enter the world with a primitive cerebral cortex, a newborn's nervous system has already spent months developing and adapting. Every single internal and external stimulus, small or large, positive or negative, has an impact. In a healthy individual, the brain takes information from past experiences and combines it with the current context to make a decision. But repeated exposure to stressors can upset this system, leading to ineffective functioning. Physiologically, this ongoing exposure to stress results in synapse (where neurons communicate) loss in the brain and affecting how the neurons branch, leading to poor connectivity and emotional dysregulation.[2] Your brain, especially the prefrontal cortex, continues developing until you are 25 years old. During that time, every internal (biological) and external (environmental) stressor and event shapes the way in which your nervous system moves you through life, making all kinds of choices *for* you.

As your brain is developing into your twenties, traumatic events strongly influence your nervous system's programming. What defines trauma is life events that are too much, too soon or too fast for our nervous system to complete a successful survival-to-safety response. In addition, trauma also takes the form of neglect and can be thought of as too little, too late or too slow when the nervous system did not receive the support, regulation and nourishment that it needed from the people it depended on (i.e. parents or primary caregivers). The encouraging news is that although the trauma that happened to you cannot be un-happened, its impact on your nervous system can be reprogrammed. As Peter Levine, the creator of somatic experiencing, says, 'Trauma is a fact of life. It does not, however, have to be a life sentence.'

Whether you experienced trauma or not, you are probably familiar with the following experiences. As a toddler, if you were scolded and yelled at every time you cried and were told *not to cry* when instead you needed to be held and soothed, then as an adult you may not have self-regulation processes in place to help you safely feel and move through difficult emotions in a healthy way. Instead, your nervous system will activate a coping mechanism (more on this shortly) by utilising well-worn neural pathways paved by the intense and frequently repeated behaviour of avoiding the overwhelming experience of abandonment and fear. This could be why you may find yourself repressing, avoiding and shutting down because your brain and body don't know what to do with big emotions. Over time, repressing, avoiding and shutting down feelings become the main ingredients in the recipes for burnout, chronic stress and illness. If in school you were ridiculed while giving a presentation and mortally embarrassed, your nervous system will record that memory and be programmed to associate public speaking with danger. Then, when you are an adult and preparing to give a presentation for work, you may find yourself anxious, breathing heavily, feeling nauseous and sweating because your nervous system is interpreting the upcoming presentation as a threat based on past experience.

Any time your nervous system perceives a threat, whether it be caused by an overdue credit-card bill, an angry rhinoceros, a pack of mean schoolmates or unresolved trauma it will deploy the stress response to survive the danger based on how it learned to do this in your developmental years. *You* may not cognitively remember the experience, *but your nervous system does.*

> **Your nervous system remembers things that you do not.**

In addition to the complexities that throw a wrench into a well-functioning nervous system, there are also coping mechanisms and survival strategies that have been honed over thousands of years. They are designed to protect you but can sometimes be incorrectly applied. You've probably heard of the 'fight or flight' response in the face of a threat or crisis. The stress response actually consists of multiple parts, including the fight, flight, freeze, appease and fawn responses (we will expand on these later). In real life, these primordial survival strategies are usually disguised and expressed through behaviour patterns as coping mechanisms. Coping mechanisms can be useful (adaptive) or unhelpful (maladaptive). Timing and duration matter as well; these adaptive mechanisms are meant to protect the body in the short term, but when used repeatedly over the long term, they can become maladaptive and counterproductive.[3]

A helpful, adaptive mechanism could be a behaviour that effectively resolves problems and leads to long-term stress reduction, such as establishing boundaries or practising self-care to achieve a better work–life balance. Conversely, an unhelpful coping mechanism may provide short-term relief but proves highly ineffective in the long run, perpetuating the problem it seeks to avoid, such as people-pleasing to escape stress, ultimately leading to even more stress. For example, through this lens, chronic anxiety is actually not an illness - chronic anxiety is a coping mechanism. It is an adaptive mechanism developed by your nervous system to tackle underlying survival needs and perceived threats that are not being adequately addressed. It is the body's way, as you've been reading, of trying to keep you alert and prepared to face potential dangers based on the map of past experiences or perceived triggers it is navigating. Ultimately, anxiety is a message coming from the body, spoken in a language you are about to learn.

We know from psychological research that explores the emotional bonds formed between children and their primary caregivers, and the studies of adverse childhood experiences (ACEs), that maladaptive coping responses can have long-term effects on a person's mental, physical and emotional health. This is particularly the case when these responses are repeatedly triggered over many years. But if you're reading this book and are healing from trauma, we want you to know that healing these long-term effects is possible and it starts with your nervous system.

So, what does all of this mean?

Whether you are experiencing messages of anxiety, burnout, illness or trauma, stress takes centre stage in the story of your nervous system because too much of it sets off a cascade of disruptions as it accumulates in your body. When stress builds up, everything can go haywire, contributing to what scientists call 'allostatic load' - a term referring to the cumulative effects of chronic stress and life events on the body. A high allostatic load strains your body's ability to regulate itself and return to a state of balance. At the cellular level, this strain brings about changes that impact on energy production, immune function, hormone signalling and cellular repair, ultimately affecting your overall health. Any imbalance in the nervous system can have widespread consequences because your nervous system plays a vital role in managing allostatic load, as it communicates with every organ, system and function in your body. Without the ability to maintain homeostasis and balance so your body can thrive - you may face persistent health issues. A review of more than 250 studies in 2020 concluded that greater allostatic load is associated with poorer health outcomes.[4] Understanding this interplay between stress, your nervous system and your body can help you make sense of why many of the other approaches you've tried have only been partially successful, and why A I R and the practices in this book can help you heal.

> **First comes stress, then come symptoms.**

Reviewing what we've covered so far, your nervous system is attempting to create feelings of safety in your body, but because of the complexities that keep it stuck and the misplaced coping mechanisms, the attempts often result in inducing anxiety and other responses that are not helpful. As a result, the nervous system makes things worse by contributing to its own cumulative load of stress, which then causes it to intensify its efforts with the same maladaptive coping mechanisms. This leads to a continuous cycle of increased stress and allostatic load. It's an exhausting feedback loop that we call *the nervous system paradox*. Put simply, in the pursuit of safety, your nervous system causes itself to become stuck and less able to actually protect and support you in regulating your health. It loses the ability to find homeostasis and move flexibly between states of activation and relaxation, and to maintain basic mental and physiological well-being. When your body is functioning in this way over a long period of time, we refer to this as living in survival mode. Eventually, this state of chronic survival mode imposes a burden on your health, increasing the stress load on your body systems and causing an increased sensitivity in your nervous system. Living in survival mode can result in chronic pain, fatigue, sleep disorders, cognitive problems, irritable bowel symptoms, addictions, compulsions and other issues.[5] Depending on what specialist you are consulting with, they may call this nervous system sensitisation, central sensitisation or dysautonomia - we call it nervous system dysregulation.

In a healthy and well-regulated nervous system, stress responses are appropriately balanced, and switched on or off based on an accurate perceived level of threat or safety. When the nervous system is dysregulated, it can result in conflicting signals being sent to different parts of the body. The key difference between a regulated and dysregulated nervous system is that a regulated system is responding to a specific, current stressor and can then return to homeostasis and balance. A dysregulated one is responding to present circumstances on the basis of past stressors (this also manifests as an unresolved *trauma response*) or anticipating ones in the future and being unable to find a way back to balance even when threats have passed. The body will continue to experience physiological adaptations to perceived threats even though it is not actually in danger. The dysregulated activation of responses creates a state of physiological imbalance, making it impossible for the body to function optimally. It contributes to feelings of stress, anxiety, overwhelm and overall discomfort, as the nervous system struggles to find equilibrium.

> **If your nervous system is stuck, then so are you.**

So, what now?

The good news is: *you don't have to stay stuck!* Intriguingly, the way out is through the same thing that got you stuck in the first place, your nervous system. It just needs your help to break out of its own feedback loop. Your body possesses an incredible capacity for adaptation and self-healing, a phenomenon known as *bioplasticity*. You may have heard of neuroplasticity, defined as 'the ability of the nervous system to change its activity in response to intrinsic or extrinsic stimuli by reorganizing its structure, functions, or connections'.[6] Essentially, the brain can reorganise itself by forming new neural connections and pathways in response to repeated new experiences and learnings. Imagine your brain's vast network of pathways is a dense forest full of trees. When you practise a new thought and learn a new skill, it's like hiking a trail through this forest. The more you repeat that exercise or learning, the more well-trodden and defined that pathway becomes, making it easier for you to access it. On the flip side, if you stop using a particular pathway, it can become overgrown and fade away, just like an unused trail in the forest. The healing practices in this book are possible because they leverage neuroplasticity - tapping the mind-body-human connection - to create new neural pathways that reinforce safety, adaptability, resilience and ease. A I R will teach you how to give your brain novel and regulating stimulation - through awareness, interruption and redesign - to draw a new map inside you, leading your body from anxious and stressed to confident and adaptive.

The goal in nervous system regulation is to rewire survival responses and maladaptive coping mechanisms and heal unresolved trauma pathways that have become our default states over time. Rewiring well-worn and automatic pathways like anxiety, depression and pain decreases the stress load on the body. And this is where bioplasticity becomes extremely significant to our healing. If anxiety, depression and pain are present in the body because the brain and nervous system are influencing our immune, endocrine and cardiovascular systems, and causing illness, then pathways that promote healing of the brain and nervous system will directly impact on all systems in the body and our overall health. In summary, the effects of chronic stress in the body can be reversed by rewiring your brain and regulating your nervous system.

A note on genetics and healing the nervous system: We know that we all inherit certain traits that make us predisposed to certain conditions *and* protected against others. While these predispositions matter, we also know through the science of epigenetics – how each gene is expressed and whether it is turned on or off – that factors like lifestyle, stress, exercise, nutrition and dozens of others have a decisive influence on whether the vulnerable predispositions in our genes or the powerful protective capacities will be realised.

For example, some individuals carry inherited genes that are associated with a significantly increased risk of cancer. But if that person is leading a healthy lifestyle, maintaining a balanced diet, engaging in regular physical activity, avoiding smoking and addressing their unhealed trauma and stress, their risk for developing cancer is reduced.[7,8] On the other hand, if someone's lifestyle diminishes their well-being, their chance of developing cancer may be enhanced.

Thus, as self-healers it is important to know that we have tremendous influence over our health. Learning the language of your body can help you address your unhealed wounds, regulate your nervous system, lower your stress levels and make informed and aligned lifestyle choices that enhance your well-being and provide an environment for you to flourish in.

You may have heard the expression 'genetics loads the gun, the environment pulls the trigger'. Though they play a role, your genes are not your fate and in addition to other lifestyle interventions, nervous system regulation is a powerful method for creating optimal health and well-being.

And believe it or not, the most capable person to make this change happen is you. You will begin to heal yourself through making the connection between your mind, your body and your human. Perhaps the most revolutionary aspect of nervous system regulation is that it is an act of self-healing. If you think about it, it's an odd thing that whenever something is wrong with us, even though it's our own mind and body, we automatically surrender our own agency and look to someone else to heal us. In some instances, this is appropriate, like when we rely on a trained surgeon to sew up a wound. But the presumption that we need someone else to heal is hugely disempowering and counter to how certain types of healing actually happen. All healers are really just assisting the patient's ability to heal themselves. Yes, the surgeon sews up the wound, but once sewed up, it's the patient's inherent ability to heal that generates new blood and skin to replace what was damaged. And yes, a therapist supports you to see things in a new light, or discover root causes of feelings, but you do the work of overcoming that root cause or making different choices to change your perspectives or reactions.

We firmly believe that you have everything you need inside you to heal. We are here to show you how to unlock that power within by communicating in the language of your nervous system and supporting your body's unparalleled ability to heal itself. As you move through the three parts of this book (mind, body, human) and apply A I R (awareness, interruption, redesign) in the exercises at the end of each chapter, you will discover the secrets your body is ready to share with you and gradually will shift away from stressed and stuck to someone who is thriving. This book will give you the guidance you need to become skilled at speaking the language of your nervous system. This book won't heal you, but you will.

And as Jen likes to say ...

You already have everything you need inside you to heal yourself.

Before we begin teaching you how to heal yourself, let's take a moment to summarise the knowledge you've acquired thus far:

- You feel stuck because there is a language your body speaks that you have yet to learn, the *language of the nervous system.*
- Your nervous system has one job: to keep you alive.
- Its efficiency in keeping you alive depends on evolutionary

algorithms which are outdated and only as good as the programming it has from your past, so instead of keeping you alive and helping you thrive, it keeps you alive and stuck in survival mode.

- Living in survival mode and in unresolved trauma patterns creates chronic stress and anxiety – creating the perfect storm for a dysregulated nervous system and illness.
- Trauma's impact on your nervous system can be reprogrammed.
- What you previously thought was happening in your mind is actually happening in your body.
- The way out is through the way in: a dysregulated nervous system is how illness is *done*, but a regulated nervous system is how illness is *undone*.

... and the best person to help you heal is *you*.

Reader's Guide

How to make this book your self-healing manual

A healer is not someone that you go to for healing. A healer is someone that triggers within you your own ability to heal yourself.
Unknown

A New Paradigm

For thousands of years, people had no idea why people got sick. Theories on what caused disease were innumerable, ranging from miasma (bad air) to evil spirits. None of these theories were as strange as they now sound, they were just the logical deductions of people trying to explain why people got sick without the knowledge of microorganisms. For example, it was observed that people exposed to rotting food, garbage or human flesh got sick and it was no big leap to presume that 'there was something bad in the air'. But, one day, a handful of pioneering (and heretical) scientists, armed with novel information-gathering tools like microscopes, began to postulate a new theory: that tiny creatures called germs - that were invisible to the naked eye - transmitted from one person to another were the cause of disease. By extension, the solution was to interrupt this transmission with hygiene and sterilisation. The application of germ theory and mass public hygiene was - and still is - the single greatest thing that scientific medicine has ever done to relieve human suffering.

Despite this revolutionary discovery and ever-increasing advances in modern medicine, the rate of diseases like anxiety, depression, chronic pain, irritable bowel syndrome, fibromyalgia, chronic fatigue syndrome, metabolic disorders, cancer, heart disease and dozens of others have been increasing at an alarming rate. Modern medicine has been struggling to successfully treat these issues because none of these disorders have their origins in bacteria or viruses - they don't fit into the *previous paradigm*.

Fortunately, we now stand at the precipice of a similar revolution in healing and relief of human suffering. A new set of pioneering (and heretical) physicians, scientists, researchers and therapists in the fields of trauma, neurology, physiology and public health, armed with new tools like fMRI and massive data sets, have postulated a new theory: *These diseases are the result of physiological adaptations rooted in a dysregulated nervous system caused by chronic stress, adverse childhood experiences and trauma*. And as such, the way to heal them is through a solution that focuses on interventions that repair, regulate and restore resilience to the nervous system. Once repaired, regulated and restored, the nervous system can reclaim its natural role as the coordinator of every organ and system in your body and the maintainer of your health and vitality.

By picking up this book, you have chosen to participate in this new revolution in healing and we are thrilled for you to experience its life-transforming and health-enhancing benefits. This is *the complete guide to regulating your nervous system and healing yourself* and we sincerely hope you will keep it close and squeeze out every drop of healing wisdom and practice that lies within. Below we will briefly review how this book will teach you to heal yourself using the new *nervous system paradigm*.

A I R

Healing yourself requires learning the secret language of your body so that you can regulate your nervous system and create lasting change. But by its very nature, your nervous system is designed to repeat the same responses to the same stimuli that have kept you safe your whole life, over and over again. So, when regulating it, it tends to resist change. Drawing from decades of research in neuroplasticity, psychology, trauma, behaviour, habit formation, our own clinical experience with thousands of clients and our personal journeys of recovering from chronic illness, we have determined that in order to make intentional, long-lasting changes to your nervous system, three ingredients are required: **awareness, interruption** and **redesign** (A I R). Awareness is recognising and understanding the responses that your nervous system is repeating. Interruption is disrupting the responses as they happen. And redesign is implementing a new response to replace the old one. Each chapter of this book is organised around teaching you a set of techniques for applying awareness, interruption and redesign to different aspects of your nervous system.

In real life A I R looks like this:

1. I notice that I am dysregulated because of a symptom (e.g. negative self-talk, anxiety, physical tension, unexplained fatigue).
2. I use an **awareness** technique like *listening* or *translating* to tune into the language of my body to get a deeper understanding of what I am experiencing and what triggered my dysregulation.
3. I use an **interruption** technique like *switching* or *modifying* to disrupt and soothe my dysregulation.

4. I use a **redesign** technique like *distancing* or *bonding* to stimulate neuroplasticity and reinforce the new regulated response to the trigger.

No nervous system is the same, so as you learn each new technique, we invite you to experiment with them and see how they work for you. Some techniques will work the very first time you try them, some may take a dozen attempts. You do not have to perfect a technique before you move on to the next one. Be creative and trust your intuition – this is *your* healing journey. A I R is not a protocol or an ultra-prescriptive method, it's the simplest and most fundamental platform to build your own nervous system regulation practice upon.

A I R best practices:

1. The laws of neuroplasticity state that the more consistently you do something the faster your brain will change and learn the new skills. For this reason, we encourage you to practise A I R techniques as often as possible and integrate them into your daily rhythm so that it doesn't feel like a task, but a healing way of life. We always say that it's not about making space for the exercises, but about moving through life in a new way that encompasses the healing tools in your everyday experience.
2. And with that in mind our invitation is not to overdo it – pressure and perfectionism tend to inhibit learning and neuroplasticity. Recognising the importance of this work for your healing and being called to practise will always work better than feeling that it's urgent and pushing yourself to do it. In fact as you do this work, you will find that perfectionism and pressure were messages coming from your body in the language of your nervous system and

when you understand the messages you will be able to shift those patterns within you.

3. The best time to practise is *in response* to a symptom like anxiety, stress, self-criticism, self-doubt, pain, procrastination or depression.

Mind, Body, Human

This book is divided into three sections representing the three realms of your nervous system and consciousness that you must develop awareness of to regulate yourself. We call them your mind, body and human. Each section is further divided into three chapters each detailing how to apply awareness, interruption and redesign to the realm being covered in that section. In mind, you will explore the cognitive and narrative realm of your nervous system. In body, you will explore the embodied and survival realm of your nervous system. And in human, you will explore the early developmental and relational realm of your nervous system. Each of these realms converse in a different dialect of the language of the body and has an enormous influence on the state of your nervous system. In learning how to listen and speak to these three realms, you will not only gain the ability to comprehensively regulate yourself, you will more comprehensively *know* yourself.

A I R	**Awareness →**	**Interruption →**	**Redesign**
Mind	listening	switching	attuning
Body	translating	modifying	settling
Human	attuning	tending	bonding

How to Use This Book

The most helpful way to use the book is in the way that suits your learning style the best. Each chapter consists of two parts: the theory and the practice. You can enjoy all the stories and science and then dive into the practices as it's been written or you can skip the science and go right into the practices. If you resonate with one section more than the other, you can immerse yourself in that one. Ultimately, we find a blend of understanding and application as well as a deep knowing of all three realms (mind, body and human) to be the most powerful. This book is the manual for your nervous system - use it as best suits you.

How This Book Can Work with Other Forms of Healing

It is our hope that this book will become your trusted manual for how to lead your own healing. The secret language of the body is a universal language and A I R is a fundamental platform that can ally with and amplify other healing modalities you are participating in. If you are doing any kind of psychotherapeutic work with a therapist, this book will help you get even more out of it and take ownership of your healing between sessions. If you are working with a physician and taking medication, this work will help your nervous system be in the best state possible to take advantage of the support that your doctor is providing you. If you are a lover of alternative or plant medicines, this book will help you become an active partner and cultivator of the healing that those complementary practices foster. In short, everything you learn in this book will act as a catalyst to every other thing you are doing to heal. And in some instances, you may find that what you accomplish by healing your nervous system will make forms of healing you've previously used no longer necessary.

A Final Note Before You Begin

All healing is an act of communication. In essence: first comes the expression of a need for help, and then comes the provision of the necessary support or cure in response to that need. This process is not just about identifying and solving a problem, but also the commitment to deep connection, understanding and resolution. In practical terms, a fever says *I have an infection* and the attuned healer can provide medicine. Bleeding says *I have a wound* and the trained surgeon can sew it up. Symptoms like anxiety, depression and irritable bowel syndrome are saying that some need is not being met and needs help. Learning the language of your body will teach you to understand what these symptoms have been saying and to provide them with what they need to resolve. Throughout this book, you will be learning this new form of communication. The illustrations below reveal what messages are really being conveyed and what self-healing can look like when these are understood. We are merely your guides; ultimately, this is your exploration of your nervous system as you make it *your healing paradigm*.

Unaware of the Language of the Body

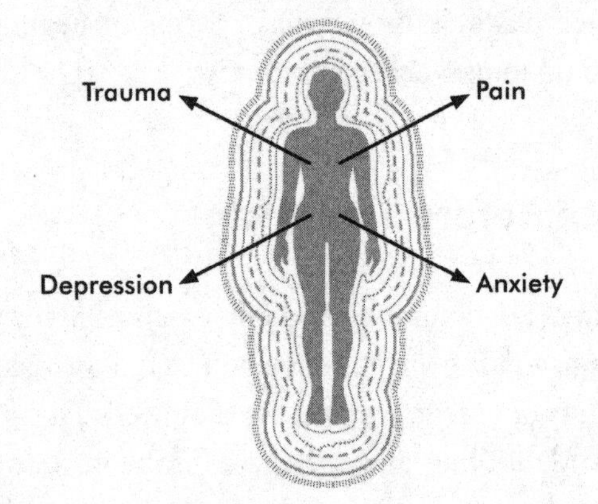

Before understanding the language of the body we are usually aware of our loudest and most alarming symptoms and their diagnostic labels.

Aware of the Language of the Body

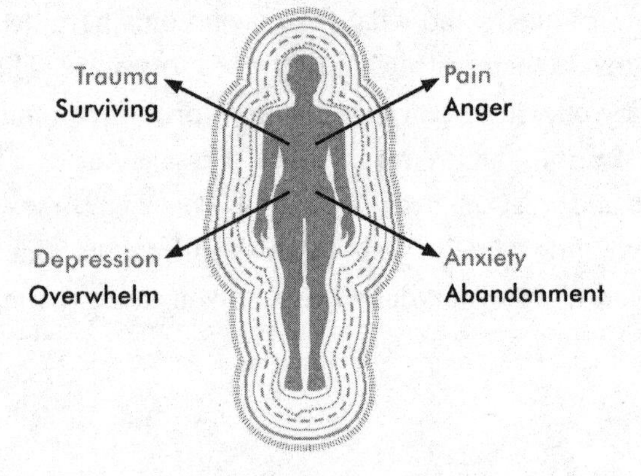

When you become aware of the language of the body, you can hear the actual messages and nuanced information underlying your symptoms.

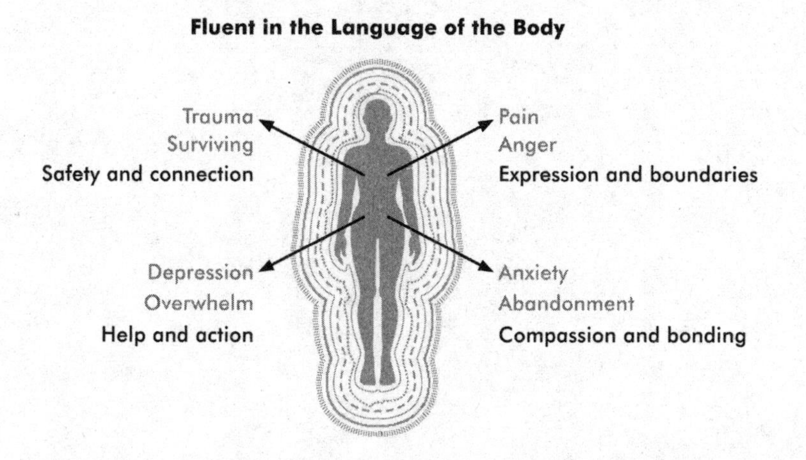

As you become fluent in the language and can communicate back to the body, you can provide your nervous system with the healing that it has been asking for.

> **This book won't heal you – you will.**

Part I
Mind

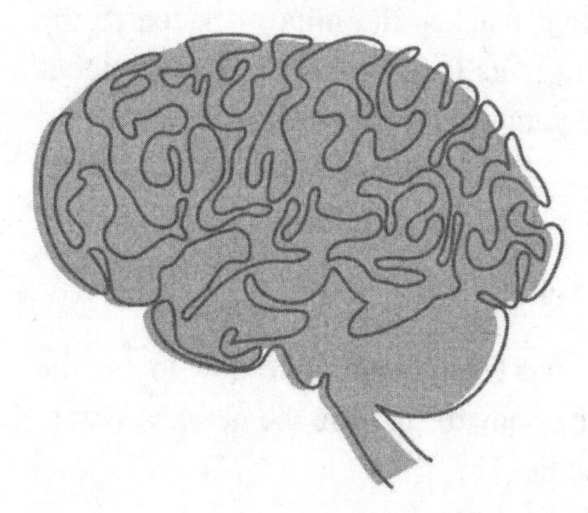

Until you make the unconscious conscious, it will direct your life and you will call it fate.
C.G. Jung

Without a mind there is no one to hear the language of the body. By using your mind, you can begin the journey of learning how to regulate your nervous system, heal yourself and transform your life.

The mind isn't just thoughts. It's where the awareness of yourself dwells and where noticing, interpreting and understanding the messages coming from your body takes place.

In the mind, you will learn how to converse with the body through the nervous system and interpret your sensations and emotions.

You will learn how to change the conversation in your mind to discover the true meaning of what is keeping you stuck, and then how to heal yourself.

The mind is where your journey into the secret language begins.

In this section, you will learn the A I R practices of the mind:

Listening

An **Awareness** technique that teaches the mind to listen to the language of the body and hear its truth, so learning more about your inner workings to heal yourself.

Switching

An **Interruption** technique that teaches the mind how to use the somatosensory system of the body to stabilise the nervous system and deepen its listening abilities.

Distancing

A **Redesign** technique that teaches the mind how to access its most transformative mode and agent for change, the observing self and allow you - for the very first time - to have a regulating dialogue with your nervous system through the language of your body.

Chapter 1: Awareness

Listening

to your nervous system through the language of your mind

Nothing in life is to be feared, it is only to be understood. Now is the time to understand more, so that we may fear less.

Marie Curie

Learning to speak the language of the nervous system actually begins with learning to listen to it. For most of us, the day-to-day experience of 'listening' to ourselves happens in our conscious mind and thoughts. As the philosopher René Descartes said: 'I think, therefore I am.' Well, we respectfully disagree and would revise his statement as: 'I think, therefore I am disassociated from the full experience of my life.' The idea behind Descartes' statement (and most people's experience of themselves) is like being in the passenger seat of a car and never realising that there was a driver's seat where you could actually drive! In other words, separating our sense of self from our bodies alienates us from the most powerful and rich aspects of who we are. As mentioned in the Introduction, the cognitive, verbal and thought-based parts of the human brain have only existed for a brief amount of time in evolutionary terms. They are also the last to form as the baby grows in the womb. In fact, those sections are housed in our neocortex, which makes up the majority of the cerebral cortex, the portion of the brain responsible for higher-level functions such as reasoning and language.[1] But beneath that 'new' brain are deeper and more fundamental brain structures that constitute the all-powerful autonomic nervous system. Unlike our neocortex, these layers have been around for a very long time and, as a result, they are the parts primarily navigating our human car through life.

These older parts of our brain house a visceral, tangible and experiential form of consciousness composed not of words but of sensations, emotions, postures, movements, energy and sense memories. This is the non-verbal form of human experience that storytellers have struggled to put into words since the dawn of time. Learning the secret language of the nervous system is actually the act of learning how to listen, communicate and influence the felt experience of sensations, emotions, postures, movements and more that will allow you eventually to regulate your nervous system. In the practice section, you will learn to listen to the messages in your body through the quality of your **breath**, the **actions** your body wants to take, the **sensations** you feel, the **emotions** you become aware of, and the thoughts, interpretations and story your **mind** has given to all of the above. For short, we call these categories of listening of breath, actions, sensations, emotions and mind BASE-M.

Throughout this book you will listen to BASE to become aware of what your nervous system is communicating to you.

When you tune into **Breath** you will become aware of:

- Locations (e.g. your belly, your chest, your lungs, your nose)
- Speed and depth (e.g. slow, fast, shallow or deep)
- Qualities (e.g. comfortable and easy or uncomfortable and restricted)

When you tune into **Action** you will become aware of:

- Posture/shape (e.g. slouched, tense, upright, relaxed or rigid)
- Potential movements (e.g. wants to fidget, jump, distract or is it stuck, still, hiding)
- Energy (e.g. high, low, sleepy, hyper, placid, buzzing)

When you tune into **Sensation** you will become aware of:

- Descriptors (e.g. tingling, tension, gripping, burning, relaxing in your body)
- Location (e.g. belly, head, chest; fixed, still, spreading or moving)
- Qualities (e.g. intensity, stuck, moving, always there, sometimes)

When you tune into **Emotion** you will become aware of:

- Labels (e.g. anger, sadness, fear)
- Location (jaw, chest, lower back, shoulders)
- Intensity (irritated or enraged, nervous or terror)
- Qualities (stuck, spreading, wants to move, blended)

Throughout the book, in the awareness step of each Part, you will be adding:

- BASE-M: Mind (thoughts),
- BASE-B: Body (survival state)
- BASE-H: Human (attachment wound)

Refer back to this page whenever you need a reminder of how to expand your awareness through BASE.

The Body Precedes the Mind by Millions of Years

To understand why so much of the brain and nervous system is embodied and influenced by things like sensations and emotions, and why our mind often gets in the way of our true felt experience, we're going to take a brief journey through a few hundred million years of brain evolution. The first organisms on earth were single-celled with a wall that separated what was inside them from what was outside them. The most primitive nervous systems arose in these cell walls, as it was at the periphery of the organism where they needed to sense and find what they could eat and avoid what could eat them. In other words, the first nervous system tissues in organisms were not brains (much less minds), but primitive sensory receptors on the surface of the organism.

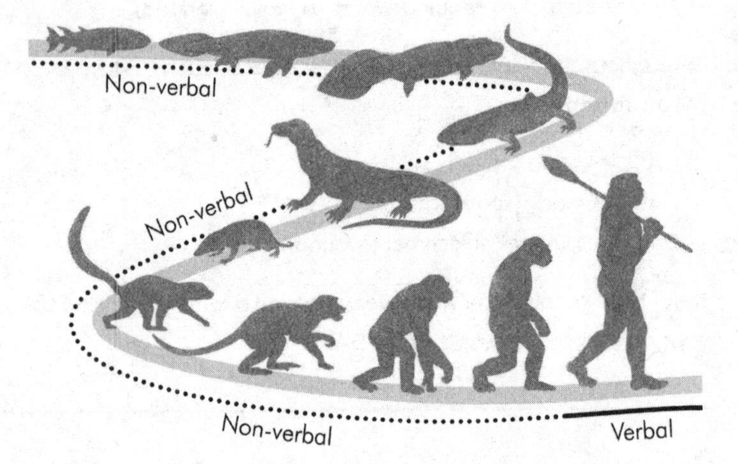

As time passed, nervous systems evolved. More and more sensory receptors came online, allowing organisms to monitor their internal and external environments. In addition, organisms developed motor neurons that allowed them to coordinate responses to stimuli (i.e. wiggle away from an amoeba that wanted to eat it). But for all intents and purposes, the nervous system was an artefact of the skin (cell wall) of the organism, and it was only after many more millions of years that organisms saw an increase in brain size relative to the rest of their body to handle much of the nervous system function. More and more neuronal tissue was concentrated into specialised sensory and processing organs (i.e. the spinal cord, brain, etc.) deeper in the body and further from the skin. This, by the way, is the biological explanation for why the skin and the nervous system arise from the same embryonic layer, the ectoderm. To be clear: the brain (processing centre) structurally and anatomically arose from the skin (feeling centre), not the other way around.

1
Notochord forms from mesoderm cells soon after gastrulation is complete
2
Signals from notochord cause inward folding of ectoderm at the neural plate
3
Ends of neural plate fuse and disconnect to form an autonomous neural tube
Notochord
Ectoderm
Neural plate
Ectoderm
Notochord
Mesoderm
Mesoderm
Neural tube
Spinal ganglion
Somite

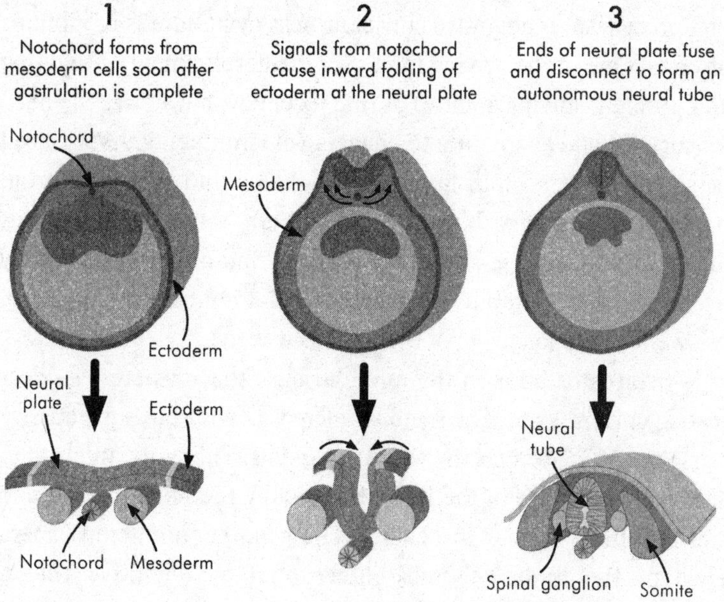

The 'skin' of the embryo 'folds' in to create the brain and central nervous system.

As tens of millions of years passed, and these one-celled organisms evolved to become organisms with brain function, the nervous system tissues that constitute embodied awareness were becoming more sophisticated, attuned to the environment and enmeshed into every corner of every organism. Therefore, it would be fairly accurate to say that the mind formed as an afterthought (pun intended) to the embodied sensorimotor-based nervous system. So, the body came *long* before the mind and the famous phrase by Descartes could more accurately be: 'I am, therefore I think.'

Despite the focus of our society and educational curricula on thinking and logic, we are creatures of feeling and movement far more often - at least as far as our brain and nervous system are concerned. The felt sense (embodiment) was the first manifestation of our nervous system and is a primary realm of consciousness that precedes our cognitive consciousness by hundreds of millions of years. One of the driving forces for the development of thinking and problem-solving abilities during recent evolution was the need for survival and adaptation to changing environments. As we faced new challenges, such as finding food, shelter and protection from predators, the ability to think, plan and innovate became increasingly advantageous ... and today we truly find ourselves in a body designed for feeling but living in a world dominated by thinking. No wonder the language of the body is a secret.

We start this book in the mind because that's where you, dear reader, currently are. The mind is where we consciously experience ourselves and interpret the world around us. Therefore, we need to speak the language of the body, but usually the mind doesn't and can become a barrier to healing. The genuine, unfiltered experience of the body becomes interrupted by cognitive filters, overthinking, old narratives and limiting beliefs. These mind-based processes often lead to selective focus, distractions, unhelpful responses and distortions in our perception of bodily sensations, feelings and reality. Excessive rumination, fear or suppression can hinder a direct and honest connection with the body's language. To break through these distortions and create a more authentic understanding of the body's true felt experience, we will be teaching you techniques that help you notice the distortions in your mind and use them as an opportunity to access your embodied consciousness, giving your mind accurate interpretations of what's going on inside you and facilitate breakthroughs in healing. The practices you will learn will directly access your neurophysiology and your most quintessential self.

> **Healing is possible when thinking and feeling happen cohesively, in the same language.**

Listening

The goal of Awareness in the mind is to teach you how to listen to the thoughts and interpretation in your mind while simultaneously penetrating through them to the unfiltered, *felt* experience of your embodied nervous system via BASE. And to access BASE, we need to introduce you to the wonderful world of interoception and the unconscious stream of information that floods your mind every waking moment. Interoception is the collection of senses and messages that come from inside your body, like knowing when you have to use the bathroom or knowing when you're hungry. Interoception is what makes listening to BASE possible. Learning to listen to your interoceptive abilities requires the same approach that learning a foreign language does: knowledge, guidance, feedback, practice, immersion and patience. But learning the language of your nervous system is *easier* than learning a foreign language because it's actually *not foreign*. It's *native* to the very core of your being. You are already immersed in it, it's always been sending messages and humans have evolved to be intimately aware of our embodied conversation. Although it may seem foreign to begin with, soon it will feel like home, we promise. Let's explore BASE and the -M of mind in more detail.

Breath

The breath is a physiological process that is managed by our autonomic nervous system but can also be directed by our conscious control, making it a perfect interface between the embodied nervous system and your conscious awareness. We will be paying attention to the breath as a part of the language of the body because listening to the quality and nature of your breathing can give you a measurement of how your nervous system is performing. If it's fast and laboured, you may be in fight or flight mode as your body wants to prime you for action. If it's shallow, slow and suspended, at times feeling like you are holding your breath, you may be in a freeze or shutdown response. Listening to your breath is an important part of becoming more aware of the mind–body–human parts of yourself.

Actions

Listening to your actions is fascinating because it allows you to notice all of the things that your nervous system 'wants to do', through your body. Actions include posture, movement (or lack thereof), expressions (like smiling), desires, impulses and habits. You may notice that you're slouching or have an urge to distract yourself by scrolling through social media or soothe yourself with salty snacks. Noticing your actions can help guide you back to the underlying nervous system state that initiated the action in the first place. For example, some people may not be well attuned to their emotional state, but they may be able to notice that their jaw is clenched. Jaw clenching is nearly always associated with the emotions of agitation and anger. So even if you are not aware of the emotion, if you notice the action, it can lead you to the underlying feelings.

Another dimension of action is energy, which takes two broad forms. The first is the more conventional sense of your energy in terms of being invigorated or tired, feeling up or down, or feeling closed or open. The second form is a more subtle concept. Every culture has a name for this subtle energy of the body but the most common labels you've likely heard before are *chi* (Chinese) and *prana* (Indian). Though most of us are not tuned into the qualities and vibrations of this subtle energy, it is intimately related to the overall state of our nervous system and can provide especially wonderful awareness of concepts that we will be covering later in the book such as regulation through movement.

Sensations

Sensations are the first point of contact between the verbal language of your mind and the somatic language of your body and connect you with the somatosensory cortex of your brain. Sensations are perhaps the richest realm of information coming from the nervous system but often the least utilised. By listening to your sensations, you are going to bring a much richer and more useful language to describe what you are actually feeling in your body. Where once you may have said 'my shoulders are tight', using A I R that statement will become something like 'my tight shoulders feel like they are literally gripping me for control'. With descriptors like that, we get closer to the truth of what your nervous system is experiencing.

Emotions

Emotions move us past the more neutral observation of sensations into the realm of feelings. Emotions can have vast meaning and depth and are the primary drivers of our *actions* (behaviour). Some impulsive emotions like fear and rage are hardwired into primitive survival parts of our brain, housed in structures like the amygdala. But as mammals whose survival (and success as a species) is dependent on our capacity to care for our offspring and the co-operation of the tribe, the more complex feelings of love, sadness, joy, guilt and shame are housed in other parts of the limbic structures of our brain.[2] The two major structures are the hippocampus and amygdala, and together they are responsible for forming memories, fear, social functions, relationships with others and our relationship with ourselves. Because emotions are such a potent communication channel of your nervous system, learning to listen to them fully can help you regulate your nervous system.

Mind

You are most familiar with the form of communication happening in the mind (your conscious and unconscious thoughts). It is the cognitive space where you process, interpret and integrate information from your senses, experiences, internal and external stimuli. Within the mind, sensory input is transformed into perceptions, patterns are recognised, memories are stored and concepts are formed. It is where you analyse, reason and draw conclusions, creating a mental representation of your reality. The mind enables you to explore abstract ideas, ponder philosophical questions, and engage in creative and critical thinking. It's the seat of consciousness and self-awareness, providing you with the ability to reflect on who you are, your place in the world and your relationship to it. In essence, the mind is the epicentre of your intellectual and emotional comprehension, making it the primary venue through which you understand and make sense of life.

As you listen, your ability to be aware of what's going on inside you will become much more sophisticated. To give you a sense of what this looks like in real life, how about a real-life example? It just so happens that while we were writing this book, Karden had a fire at his co-working space that he owns and works from. Shortly after the fire, he got a message from his insurance company saying that they were terminating his coverage for no reason other than he had filed a claim for help. At the level of *mind*, Karden thought, 'This is unfair and concerning.' But what else was going on? When he slowed down to listen to the messages of his body, there was *a lot* going on. His *breath* was shallow and accelerated, he could feel anxious energy in his body that wanted to take the *action* of scrolling through his phone to find the number of his insurance agent to help him solve the problem, he had *sensations* of feeling trapped and constricted in his chest, and he had *emotions* of anger and fear. Karden's mind-level perception was 'this is unfair and inconvenient', but from the perspective of his embodied nervous system (BASE), there was a threat and all systems were being called upon to survive it.

When you learn to listen, there is a lot being communicated in what Peter Levine, a pioneering trauma researcher and founder of somatic experiencing, calls an 'unspoken voice'. Hearing it all takes time but with patience it will develop like any other skill you've wanted to learn. Let's imagine gardening. Before someone becomes interested in the world of plants, when they look at a garden (BASE), all they see is an amalgam of green *stuff*. But after a few days or weeks of time spent in the soil, their eyes and brain begin to discern various shades of green, shapes of leaves, styles of stems and individual plants like lettuce, beans, squash and potatoes. What was once a sea of green now transforms into a vibrant space brimming with different species of flora and countless intricate details. Moreover, your ability to understand (BASE-M) what's going on in the garden and your ability to care for it improves. For example, before you learned about gardening, if you had a plant that wasn't growing well, your mind (with its limited information) might guess it needed more water but by giving it more water the plant got worse. But with your new knowledge of gardening earned by a few weeks spent in the soil, you can go beyond guesswork and observe that the plant is actually suffering from a fungus, and in fact needs more sun to dry out, not more water. The most fascinating part is that once the brain attunes itself to notice these subtleties of the mind and body through listening and awareness, it retains the ability to perceive the richness and depth of the garden forever. As you develop the ability to skilfully listen to the interoceptive information flowing from your body to your conscious awareness, you will be able to use your mind as an observant and skilful gardener to heal your nervous system.

Breath – where embodiment and consciousness meet
Action – the movements of your body driven by your brain
Sensation – the raw input coming from your body
Emotion – the sophisticated input coming from your body and interpreted by the mind

Mind – where meaning is made
Body – where everything is felt
Human – the very thread that weaves all of you together

The Meaning-Making Mind

The mind, while an extraordinary instrument of understanding, can also become a formidable obstacle when it gets trapped in cycles of misinterpretation, distraction, analysis, overthinking and catastrophising. When it doesn't know how to shift out of these processes, it can keep us stuck in states of anxiety, depression and overwhelm by reinforcing the dysregulated states in the body with unhelpful and inaccurate thoughts. You may think this is only an experience of the mind. But what is actually happening is that the mind is getting stuck in unhelpful thinking and belief patterns as a result of the unfelt, unprocessed and unhealed experiences in your body that it doesn't understand. Until the mind learns to listen to its internal narrative *and* what the body is actually communicating, you won't be able to make sense of what's actually going on inside you. For example, when you have anxiety, the mind engages in an unsettling dance with the body by trying to interpret the bodily sensations that feel uncomfortable, intense and threatening. It searches for meaning in the pounding heart, the shallow breaths and the knots in the stomach, attempting to construct a narrative that can rationalise these visceral responses. This innate drive to interpret the unprocessed experience of the body stems from our primal instincts for self-preservation. It seeks to identify potential threats and mobilise a response to find the experience of safety. Unfortunately, since it doesn't know what's actually happening, not only can it not solve the problem, it usually leads to negative thought patterns and inaccurate assessments like 'I must be dying' that send messages back into the body that amplify and reinforce the dysregulation.

Learning the secret language of the body begins with learning to listen, without judgement, to the unhelpful and habitual thoughts and narrative that are going on in your mind and peer beneath them to the experience in your body. Thus, becoming aware of your thoughts and narrative through the BASE-M framework becomes the portal to discovering the root causes behind the chronic symptoms you have been unable to resolve. Put another way, if your awareness notices that you have the thought 'I have so much to do, I'm not going to get it done and I'm going to disappoint everyone' in your mind (-M), that is the cue for you to pause and expand your awareness into BASE so you can *listen* to the language of your body and learn what's really going on in your nervous system. Our work with Roy illustrates this process in action.

Roy, *a 47-year-old CEO experiencing chronic migraines, chronic low back pain and anxiety, came to us for help. He was frustrated and angry at his symptoms and we – knowing that chronic symptoms rarely communicate the obvious – immediately sensed an underlying reason that wasn't on Roy's radar. Roy heard about our work with the nervous system and requested a 'biohacking fix' because he didn't have the time for anything else. We told him we could help (without fully disclosing that there would be no 'biohacking fix'). He asked us to give him some exercises to help his back and headaches because he was spending too much time and money on massages that weren't fixing them. Instead we asked him why he thought he was experiencing these symptoms. Roy replied, 'I think it's because I don't exercise enough.' We asked him why he didn't exercise enough and Roy replied, 'Because work takes up too much of my life, and I'm tired and frustrated.' We asked Roy what he felt about his job and he said, 'I don't love it or the people I work with, and these headaches and back pain don't make it any better.' Roy thought that he needed biohacking tools to fix his head and back so that he could go back to tolerating his job and managing people, but we guided him in a different direction.*

Since his conflict around work was palpable to us, we looked at the Mind component (from BASE-M) and coached him to notice his thoughts around work. And what came up for him was frustration, anger and dislike towards his team. With this in mind, we guided him into noticing his breath, actions, sensations and emotions (BASE). He reported restricted breathing, that his body wanted to move but felt stuck, that he had tension, and pressure in his neck and shoulders, and he was quite surprised to feel that underneath his initial irritation were the emotions of overwhelm and fear in his chest. The story in his mind did not reflect the very intense fear-based messages being communicated by his body.

Next we guided him to notice the thoughts coming up, except this time we encouraged him to not base the thoughts on the story he was already familiar with, but instead to base them on the feelings in his body. This time, what came up for him was, 'I'm afraid that people at work will reject me if I say or do something wrong and that feeling overwhelms me to the point where I feel like I want to run away and hide, but I can't. And as I'm saying this my headache is really pounding and my back is really hurting.' Looking at how Roy's mind interpreted the sensations in his body versus what the sensations were really communicating, we could tell that there was a huge gap in his own understanding of what was going on for him. What Roy thought was anger and frustration was actually hurt and fear. What Roy thought was disliking people was actually fear of them disliking and rejecting him. Roy was shocked that turning his attention to the fear lessened the intensity of his symptoms – as if the symptoms were finally being heard and wanted to communicate this.

Learning the true messages coming from his body helped guide Roy's healing. By practising A I R and diving deeply into this mind–body–human experience of himself, Roy learned that his childhood wounds caused him to be insecure in his relationships and hyper-

vigilant toward people's behaviours as an adult. Roy's migraines and back pain were not caused by lack of exercise, but by the perpetual sense of fear and deeply rooted insecurity within himself. This awareness was the first step that opened Roy to the world of nervous system regulation and healing. Roy has fully recovered from migraines and back pain and is still a CEO. There was no biohacking fix, but there was an incredible amount of self-healing and repair within his mind–body–human.

> **Your mind invents meaning to make sense of the experience in the body and until it learns what the body is actually communicating, it can get stuck in a story that isn't true.**

Listening Is Medicine

Listening is much more than just hearing the messages from your nervous system; the skill of listening for awareness is a deeply regulating practice in and of itself. We'll get to this in more detail later in the book but, essentially, being listened to is powerful medicine for your mind and body. You already know this from your everyday life. We all long for connection and to be understood. In fact, it's hardwired into our nervous systems as mammals to seek safety and regulation from our fellow humans. This is called co-regulation. Listening practices tap into this need and into the healing power of co-regulation by directing the observing part of us to *listen to ourselves*. In doing so, we can be our mind and body's own best friend during times of tension or crisis. But it is important to remember, listening is about being present and paying attention to *what* is happening, not *why* it's happening.

This distinction is critical to remember. Asking 'what' activates curiosity. Curiosity is the magic ingredient that drives learning and understanding. When you are curious, you are primed to be in a regulating state for your nervous system. Curiosity and observing *what* is happening is a very distinct action from asking *why*. Asking why (at least too early in the listening process) tends to take people out of their interoceptive experience and into their cognitive story, which is the opposite of what we are trying to do with listening. For most of us, the cognitive act of engaging with *why* is blended with judgement and limiting beliefs that are already ingrained in our brain. As you might imagine, judgement is *not* a safe regulating state for the nervous system. So 'be curious, not judgemental' helps us to observe and understand what is happening before jumping to why it's happening and analysing all of the reasons, getting stuck in the meaning-making machine that is the mind. The moment we shift into thinking about why, we've stopped listening. The moment we recognise this, listening becomes medicine. Here is an example of how the story of the mind can spiral out of control with *why* and lead to a pattern of constant health anxiety.

Jake *came to us because of anxiety and symptoms of increased heart rate and difficulty breathing when triggered and was determined to find out why. Jake had been to doctors and specialists to rule out any serious condition, but despite being told there was nothing physically wrong with him, they also didn't know how to help him. We asked him to describe how he gets triggered, what he thinks and what he feels, and he said: 'I don't really know what triggers me. All of a sudden I can feel my heart rate quickening and my chest getting so tight that I can't breathe ... then I get scared and think: "Why is this happening? What if I can't breathe? Will I faint? I know the doctor said it's anxiety, but what if there is something else going on? What if the doctors missed something? What if there is* *something wrong with me?" and I start to freak out even more, it's awful, and sometimes I get a panic attack.'*

trigger → uncomfortable feelings in the body (hurt, fear, anxiety) → negative self-talk → uncomfortable feelings in the body get worse → body becomes stressed → symptoms begin (heart and chest) → mind catastrophises → symptoms get worse → panic attack

In Jake's case, his mind's quest for why and his reaction to his body's uncomfortable body sensations was extremely unhelpful and amplified the dysregulation in his nervous system and by extension his physical symptoms until both his mind and body got so activated that he would have a panic attack. Moreover, Jake's inability to listen skilfully to his mind and body was also blocking him from identifying the trigger that began the response in the first place. We encouraged Jake to be curious, like a detective, and asked Jake to think back to his most recent episode and tell us what happened beforehand. He told us he had been at the gym, so while coaching him to pay attention to BASE, we asked him to see if he could recollect anything else that happened while he was there. After about a minute of feeling and pondering he said: 'There's a girl that I like who exercises at my gym. And, come to think of it, on that day I saw her. We made eye contact and I smiled at her and she looked away and didn't smile back.' This piqued our curiosity so we encouraged him to stay with the sensations in his memory and asked him if anything else happened. He said: 'At the time, I brushed over it but now I can feel that it made me anxious and it made me feel a lot of self-doubt.' We asked Jake to tune into the thoughts he may have been thinking in the moments after she looked away. 'She doesn't like me. Did I offend her? Is something wrong with me? I'm such a failure.' As he shared these thoughts he recalled, he said he could feel his heartbeat become loud and fast and his chest feel tight. We asked him if that was OK, and he said: 'Yes, it doesn't feel good, but it's OK. I'm just amazed that that girl triggered me so much.'

> Listening → curiosity over judgement (what's happening instead of why it's happening) → a sense of safety → no symptoms → breakthrough discovery → understanding what is actually at the root of the trigger.

As it became clear, his mind and body had been triggered before the symptoms got loud enough for Jake to notice. The initial feeling of rejection in his body at the gym had initiated a conversation of low self-esteem and self-criticism in his mind which started a dysregulating feedback loop between his mind and body. Jake missed the thought-based response to the trigger of being ignored. Then later when his symptoms increased, he got trapped in his health anxiety coping mechanism as his mind tried to make meaning and fix the threatening experience he was having in his body. Like many, Jake is someone who never learned to feel feelings and emotions, and since he couldn't feel them – but could feel the physical experiences – he became hyper-fixated on thoughts and symptoms.

Jake was surprised when he realised that the interaction he had with the girl at the gym, which he barely recollected, had been the actual trigger. The first skill we taught him was to listen to BASE-M anytime he was triggered instead of getting caught up in his health anxiety dialogue. A week later he reported amazing improvement. Not a single panic attack all week. He said that sometimes he was doing BASE-M a few times an hour! Each time he'd notice something uncomfortable in his body or anxious thoughts in his mind, he would say out loud, 'What am I experiencing?' He would then listen to BASE and name the feelings out loud and say: 'I am feeling the beat of my heart. I am feeling a tightening in my chest. I feel like I want to run and hide. I am feeling fear. These feelings are OK. They may be uncomfortable, but I am not in danger. I am here to listen to my feelings.' He said that although he was shocked at how often he was triggered, every time he listened to himself in the language of his nervous system, and made space for his feelings, he felt better. The fact that he could stop his breath and heart from spiralling into anxiety and panic, simply by being with and labelling BASE-M, felt like he had superpowers!

Jake's experience not only demonstrates the healing power of *listening*, it also demonstrates that every cell in our body is listening to and reacting to our thoughts. Research has shown that our mental states can influence the physiological processes within our cells. When we experience stress, anxiety or even joy, the brain sends signals to the body that trigger the release of various neurochemicals and hormones. These biochemical messengers then impact on our physiology, *literally*, influencing our health. For example, researchers at Yale wanted to see if a person's mindset could impact on the physiology of being full. On two separate occasions, they told participants they were getting either a 620-calorie 'indulgent' shake, or a 140-calorie 'sensible' shake. Except - in reality, they were all getting the exact same 380-calorie milkshake. The findings were fascinating. When participants thought they were being indulgent, their bodies produced a dramatically steeper decline in the 'hunger hormone' released by the stomach after drinking the shake.[3] After they thought they had the sensible shake, there was a flat hormone response. In other words, simply thinking they were having an indulgent milkshake not only made them feel fuller faster, their brain caused their body to literally produced *more* satiation chemicals. If just approaching food in a different way can change your physiology, imagine what transforming the conversation in your mind so that it regulates your entire nervous system can do for your well-being!

Generally, you've probably heard a description of this science as the mind-body connection. Jo Marchant writes about it in her book *Cure*:

> We no longer need to abandon evidence and rational thinking in order to benefit from the curative properties of the mind. The science is there ... despite their best intentions, medical professionals are working within a system that prioritizes access to medical technology and allows increasingly little space for the human aspects of care ... this paradigm has been less successful at warding off complex problems such as pain and depression or stemming the rise of chronic conditions such as heart disease, diabetes and dementia.[4]

In this profound exploration of the mind-body-human connection, you will uncover the truth that your mind can actually be finely tuned to the authentic signals coming from your body. In turn, your mind will gain transformative clarity which alters how you perceive yourself, your thoughts and the experience of your life. With listening, you will begin unlocking the secret language of the body and by doing so gain insight into the real forces driving your reactions, behaviours, symptoms and patterns.

> **Until your mind is able to shift out of its habitual reactionary patterns and learn the truth in the body – you will stay stuck.**

Let's summarise what you've learned so far about Listening for Awareness:

- We are feeling beings before we are thinking beings.
- Tuning into the interoceptive signals from your body using breath, actions, sensations and emotions BASE-M is the first step in learning to speak the *language of the nervous system*.
- The mind is usually misinterpreting the experience of the body and your unhelpful thoughts are cues to start listening to BASE to learn what is really going on.
- The conversation in your mind has a powerful influence on whether you are regulated or dysregulated.
- We can help ourselves through active listening for awareness and calm our mind's and our body's reactions to stressors.
- Listening without judgement helps you move from critical thoughts to a healing mindset.

Practices 1

Awareness: Listening

Teach your mind to listen to what the body is actually communicating

In these practices we are going to guide you in directing your attention to BASE-M in a variety of ways. You will learn to notice what your body is actually feeling and trying to communicate in relation to your thoughts and perceptions.

Why Do It?

As you learn to listen to the language of your body you will be able to decipher what is going on in/with your nervous system. With this information you will be ready to start influencing the nervous system states that are actually driving your anxiety, stress and symptoms as well as your unhelpful thoughts and perceptions.

When to Do It

Use listening anytime you detect that you are in unhelpful thought patterns and behaviour. Since this practice is primarily about expanding your awareness of what's happening in your nervous system, we suggest you practise it as frequently as possible. The more you make the choice to notice BASE-M when you catch yourself in unhelpful thoughts, the faster you will train your mind and body to address the reality of what's happening inside you.

Tips before you begin

- Your brain is a neural network which is built around billions of associations. Therefore, memories can be strongly associated with other memories. Though there is nothing wrong with this and it can be a powerful tool for insight, it may be disruptive when trying to learn the basics of listening for awareness. For example, if you pick a memory of a pet for the positive contrast practice (below) but that pet is no longer alive, it might swing you into the negative experience of sadness and grief. We invite you to pick **mildly** stimulating and ordinary experiences that are less likely to distract you from learning contrast and feeling its intended effects.
- No nervous system state is inherently positive or negative. The sadness mentioned above at the loss of a pet is not 'negative', but for the purposes of this exercise we are using the terms positive and negative for simplicity.

Practice 1: Contrast Exercise

Positive Contrast

1. Find a comfortable place to sit.
2. Think about a mildly positive experience, a real or imagined event or memory that sparks positivity. For example:

discovering a beautiful sunset, receiving a compliment, listening to uplifting music, receiving a gift, walking on a beach ...

3. While you tune into this positive experience, begin listening to BASE-M.
    a. **Breath:** What am I feeling in my breath?
    E.g. *I am feeling an opening of my lungs, a softening in the back of my throat as I inhale, a sense of not needing to inhale deeply*
    b. **Action:** What does my body want to do and how does it want to move?
    E.g. *My body wants to expand, open itself, bask, I want to spread my arms, my body wants to soften, smile and skip*
    c. **Sensation:** What sensation am I feeling in my body and where?
    E.g. *I am feeling warmth in my chest and belly, tingling in my hands, calm*
    d. **Emotion:** What emotion am I feeling and where?
    E.g. *I am feeling happiness in my face and mouth, I am feeling gratitude where my heart is*
    e. **Mind:** What thoughts am I having?
    E.g. *'This is nice.'*

Negative Contrast

1. Find a comfortable place to sit.
2. Think about a mildly negative experience, a real or imagined event or memory that sparks negativity. For example: missing public transportation, sitting in traffic, misplacing keys, flat tyre, spilled coffee on your new clothes, rain at a picnic.

3. While you tune into this negative experience, begin listening to your body:
   a. **Breath**: What am I feeling in my breath?
      E.g. *I am feeling my breath quicken, I felt like I need to inhale deeply, my lungs feel tighter than before*
   b. **Action**: What does my body want to do and how does it want to move?
      E.g. *My body wants to constrict, close off, take a strong stance, my body wants to stomp my feet, 'fix it'*
   c. **Sensation**: What am I feeling in my body and where?
      E.g. *I am feeling warmth in my legs, I am feeling a flushing hot sensation in my face, I am feeling my heart rate quicken*
   d. **Emotion**: How am I feeling and where?
      E.g. *I am feeling annoyance in my chest, I am feeling frustration in my jaw, I am feeling disappointment in my shoulders, perhaps fear*
   e. **Mind**: What thoughts am I having?
      E.g. *'This is annoying and frustrating'*

Practice 2: Noticing Your Mind's Story

Use your thoughts as cues to tune into BASE-M and catch yourself having unhelpful thoughts, for example if you notice yourself saying things like: *I can't do anything right, Nobody loves me, I am a failure, I can't change, I will never recover, I'm stuck, I'm broken ...*

1. When you notice an unhelpful thought.
2. Pause.
3. Take a deep breath.

4. Focus your attention on the following. Instead of trying to change anything about these elements, allow yourself to simply observe and label them as they are:
    a. Breath for 1 minute, describe it
    b. Action for 1 minute, describe it
    c. Sensation for 1 minute, describe it
    d. Emotion for 1 minute, describe it
5. Take note of what's going on in breath, action, sensation and emotion and observe how the experience of your body is underlying the thought pattern you are having.
6. Describe to yourself how you feel now and if anything has changed.

PS: You may find that by noticing your body and labelling BASE, that your mind-body actually feels better and the intensity of the unhelpful thought pattern decreases.

Make Listening a Part of Your Life

The goal is to make your mind a helpful participant in regulating your nervous system. To do this, we invite you to:

1. Practise listening (tuning into BASE-M) as part of your everyday life.

2. Commit yourself to listening to your body whenever you have a disruptive thought or notice dysregulation.
3. Gradually create a new conception of your mind–body where your body's messages are as important as the activity of your mind.

In the beginning, listening takes time as you become familiar with all your body has to say and the story you are observing with curiosity. Over time your practice will shift to a swifter, natural and intuitive check-in.

PS: By practising listening regularly, you are subtly *interrupting* your default patterns of nervous system dysregulation. By consciously choosing to notice and be curious about what's going on in the body rather than allowing your mind to dominate with its own meaning, you are sending a powerful message to your nervous system that it's time to do things differently.

Notes:

Chapter 2: Interruption

Switching

from unhelpful thoughts

When you're losing your mind, find your body.

Karden Rabin

Listening, the first awareness practice in A I R, prompts us to observe our mind while simultaneously tuning into the true messages coming from our bodies that we need to hear to heal ourselves. As you are learning, becoming fluent in the secret language of your body requires learning how to become aware of your body's experience as opposed to relying only on your mind and its often inaccurate interpretations, as we have just explored in the previous chapter. But trying to use your mind to listen while it itself is confused, overthinking and dysregulated can be challenging. Switching, the first interruption practice in A I R, helps you stop the unhelpful thoughts and conversations in the mind by refocusing your attention and actions into the experience of your body through the nervous system's super highway of perception: the *somatosensory system*. Switching has the power to simultaneously shift you out of unhelpful thought processes and survival responses and activate regulating physical pathways that soothe the mind in such a way that it can become a more helpful, more accurate and more compassionate interpreter of the signals coming from your body. By the end of this chapter, you will have an amazing collection of somatosensory tools that will not only help make your mind a much more helpful place, but also make you feel more connected, whole and aware of your mind-body-human.

The Somatosensory System

Bessel van der Kolk, in *The Body Keeps the Score*, says that 'one of the clearest lessons from contemporary neuroscience is that our sense of ourselves is anchored in a vital connection with our bodies'. Long before neuroscience proved this, the pioneers that established the field of somatics in the early- and mid-twentieth century knew this to be true. Thomas Hanna, who coined the term somatics said, 'the human body is not an instrument to be used, but a realm of one's being to be experienced, explored, enriched and, thereby, educated.' And it's the principles of embodied awareness and movement developed by Hanna as well as Wilhelm Reich, Moshe Feldenkrais, Emilie Conrad, Frederick Alexander and many more, that make up the bedrock of contemporary nervous system regulation practices and trauma therapies. Therefore, you will be seeing these mind-body principles integrated into some of the later practices. But the bottom line is that the more connected and comfortable you are in the embodied aspects of your nervous system, the more stable the psychological (mind) aspects will be.

Somatics (embodied practices) are possible because of our somatosensory system. Anatomically speaking, the somatosensory system encompasses a complex network of structures and pathways that facilitate the perception and interpretation of various sensory stimuli originating from the body's surface, muscles, joints and internal organs. These networks contribute to helping your mind and body 'know' its shape, position and boundaries through monitoring and interpreting your external environment and your internal physiological state. This system is crucial for your ability to perceive and respond to tactile sensations, temperature changes, pain and movement. At its core, the somatosensory system consists of specialised sensory nerve receptors distributed throughout the body, each attuned to specific types of sensory information. This sensation information can be divided into three main categories: interoception (internal physiological stimuli), proprioception (the sense of where the body is in space) and exteroception (external stimuli perceived in our environment). We have already learned about some of the interoception pathways through BASE, and in this chapter we will be sharing more about proprioception and exteroception (the five senses) and how these can help you self-heal.

Proprioception

While interoception covers the internal messages your own body is communicating to you, proprioception monitors where your body is in space along with speed and length of movement. It is through proprioception that you know that your arms are above your legs or, if you are in a handstand, that your arms are temporarily below your legs. It is through proprioception that your brain monitors the highly skilled physical movements needed to do something like surgery or a ballet pirouette. And it is through proprioception that you know if a stretch is putting too much stress on a joint, nudging you to back off. Proprioception is the real-time GPS of your mind-body, constantly transmitting information.

In our cars, GPS allows us to clearly and confidently navigate to our destination. We may feel anxiety if we are travelling and our GPS stops working, but we may also feel an instant wave of relief when it comes back online. If we spend all of our time and attention in our minds, we lose this important reception with the body. Switching can help us tune into our proprioceptive awareness for feedback that gives our brain a sense of clarity and confidence through knowing where we are in space. In addition, it gives another dimension of definition and reality to *being in our body*, grounding us in our physical selves and giving us a sense of where we are relative to the world around us.

But proprioception is so much more than just getting a GPS location on yourself. Jen used to be a professional ballet dancer, training for the stage while unknowingly honing her proprioceptive sense. Through countless hours of practice from a young age, her body became a finely tuned instrument, responding effortlessly to her every direction and intention. Movement, for Jen, was not just a physical endeavour; it was a helpful escape during times when her unresolved trauma felt heavy and kept her stuck. It became a gateway to a mental state that felt safer, more whole and calm. In the midst of pliés, pirouettes and arabesques, she often found herself instantly feeling more grounded. In this calmer state, Jen's mind was able to disengage from fear and anxiety (her most persistent patterns), allowing her to feel more present.

Years after her ballet career, Jen can still harness the power of fine movements, whether through dance, qigong or yoga, to nurture her mind and spirit as a supportive regulator for her nervous system. Engaging in these practices still, today, offers her a sanctuary of self-care, a space where she can quiet the noise of her mind, come back to her body and attune herself to the whispers of her nervous system. It is really through the connection to her proprioceptive sense that Jen found her way to healing and the healing work in this book when she was chronically ill (you'll learn more about this later on in the book). Research on proprioceptive awareness via yoga done by Bessel van der Kolk supports Jen's experience. His findings demonstrated that the areas of the brain responsible for embodied self-awareness (the insula) and executive function (the prefrontal cortex) increased in activity after 20 weeks of yoga practice, a highly proprioceptive and touch-based experience. In fact, findings from other researchers support this as well. A review of 11 different studies examining the effects of yoga on the brain concluded that it has a positive effect on the structure or function of the hippocampus, amygdala and prefrontal cortex.[1]

Exteroception (the Five Senses)

Sight, smell, hearing, taste and touch are the elements of the nervous system primarily concerned with monitoring and responding to your external environment. Because of the critical nature of the information supplied by these senses, each one (other than touch) has an individual cranial nerve associated with it to transmit signals directly into deep brain structures. They are then perceived by corresponding parts of the neocortex: olfactory, gustatory, visual, auditory and somatosensory cortices. Although in modern times our five senses may be less critical to our survival than they were for our ancestors, as far as your autonomic nervous system is concerned, they are still just as important to its perception of safety now as when we were hunting and gathering in the savanna.

When we use switching techniques to focus our attention on our senses, our nervous system tends to respond in a highly regulated way. This is because it gets to do what it was designed to do, monitor the environment and determine whether you are safe or not. Typically, if you are not literally in a threatening environment, when you actually listen with your ears, your nervous system notices that there aren't any dangerous sounds. In fact, maybe there are some pleasant noises, like the singing of birds. When you truly see with your eyes, your nervous system notices that there aren't any wolves or lions, instead there are some pleasant items to see, like a vase of flowers on the table. And when you inhale deeply with your nose to smell, your nervous system notices that not only are there no toxic scents but some pleasant ones, like the smell of lavender in the moisturiser on your skin. This active gathering of positive information from your environment using your exteroceptive senses instantly gives positive feedback to your nervous system and helps it to regulate and calm. Too often, we are all running through life so quickly that our senses aren't able to gather this information and feedback in a meaningful way. This is the reason behind the phrase 'stop and smell the roses'.

Karden learned the regulating power of tuning into the five senses many years ago during his time teaching wellness seminars for veterans with the Wounded Warrior Project. As often happens with many powerful learnings, it was preceded first by making a mistake. Karden had attempted to lead the veterans through a traditional closed-eyed mindfulness meditation practice. To his surprise, within just a few minutes of sitting still with their eyes closed, most of them became highly *dysregulated*, with some of the soldiers and marines getting so activated that they had to leave the room. Because every member of this group had post-traumatic stress disorder (PTSD) from their tours in Iraq and Afghanistan, sitting still with their eyes closed made them feel incredibly vulnerable to attack. Karden felt awful about his mistake but was determined to find a way to support the veterans in finding more peace, so he decided to try a different approach. The next day, as the group climbed a peak together in the Catskill Mountains, Karden asked each of the veterans to find comfortable places to sit and then led an open-eyed, five-senses immersion practice. The results were the complete opposite of the day before. Many of the veterans reported that the 10-minute exercise was the most profound experience of tranquillity that they had experienced since leaving military service.

The reason for the radically different outcomes in these scenarios is that the five-senses immersion supported their nervous systems in doing its primary job, as it was able to maintain constant surveillance of its surroundings, keeping them safe and alive. As each sense reported back to the brain that the environment was safe, the brain sent messages to their bodies that they could relax. One sense at a time, each service member was able to let their guard down a bit more. In trauma-informed somatic therapies, this process of allowing the brain to take in and assess its environment is called orienting. You too will have a similar experience as you use switching techniques to orient yourself and get out of your mental-based perceptive networks and into the body-based perceptive networks to become more fluent in the full mind–body–human connection. Along with interoception, these exteroceptive networks give you direct access to the secret language of the nervous system.

> **Where your attention goes energy flows.**

Touch

Touch is also the 'felt sense' or 'perception of the body', and is a term that not only encompasses the traditional sense of touch but also the sensations coming from our body that we discussed in Chapter 1. But it is so much more than that. Touching skin-to-skin is how first contact between mother and child happens after birth and it continues to be the most important form of connection and communication through early childhood development. Research on touch has shown that touch is 'fundamental to human connection, bonding, and health'.[2] Touch is one of the most profound ways in which we can really *know who we are* and experience what it's like *to be at home in ourselves*. To further illuminate this concept, we are going to borrow generously for a moment from Deane Juhan, author of *Job's Body*:

> This close association between the skin [where touch happens] and the central nervous system could not have more concrete anatomical and physiological connections. Skin and brain develop from exactly the same primitive cells. Depending upon how you look at it, the skin is the outer surface of the brain, or the brain is the deepest layer of the skin. Surface and innermost core spring from the same mother tissue, and throughout the life of the organism they function as a single unit ... My tactile experience is just as central to my thought processes as are language skills or categories of logic. The skin is no more separated from the brain than the surface of a lake is separate from its depths; the two are different locations in a continuous medium. 'Peripheral' and 'central' are merely spatial distinctions, distinctions which do more harm than good if they lure us into forgetting that the brain is a single functional unit, from cortex to fingertips to toes. To touch the surface is to stir the depths.[3]

As Deane says, and as you learned in Chapter 1, the brain and skin are simply two different locations in a *continuous* medium. Reconnecting to the felt sense with switch practices transports you from the thinking mind location of the nervous system, often unhelpful and detrimental when we are dysregulated, to the part of our nervous system that provides us with our *very own sense of self*.

How Switching Works

The sensory receptors of your nervous system that are responsible for perceiving body position, movement and your five senses will be stimulated through a series of exercises like pushing your arms against a wall, balancing on one leg while executing a coordination task or self-massaging your feet or arms. These exercises leverage sensory and motor networks that plug you into embodied awareness and pivot you out of the unhelpful cognitive processes that have been trying to keep you safe but have been keeping you stuck. In 2016, Tal Shafir, an interdisciplinary researcher specialising in movement–emotion interaction and its underlying brain mechanisms, found that 'through deliberate control of motor behavior and its consequent proprioception and interoception, one could regulate his emotions and affect his feelings'.[4] Antonio Damasio, a neuroscientist and neurologist, and the founder of the theory of emotions, found that emotions are generated by conveying the current state of the body to the brain through interoception (input from the physiological signals of the body) *and* proprioception (input from muscles and joints). What this means is that you can influence the state of your mind through both the feelings in your body *and* the movement and sensory experience of it. You've likely experienced being in a bad mood until a friend turns on your favourite song prompting you to dance and 'let off some steam', then suddenly moments later you're smiling and feeling better. Listening to music together while moving to it conveys an immense amount of information to your nervous system, giving your brain a great amount of new information to organise that is different from the thoughts that have been keeping you stuck.

There are two senses in particular that have a profound effect on your nervous system through movement: the vestibular sense, controlling balance and movement; and the proprioceptive sense, controlling body awareness. By engaging in specific movements, postures and practices, the practice of *switching* will help you activate the somatosensory system, creating a two-fold effect. Firstly, this sensory input triggers neural responses that directly influence your brain's emotional regulation centres, prompting a shift away from survival and 'fight or flight' states. Secondly, the activation of proprioceptors in your body and the five senses prompts a curious and receptive sense of bodily awareness in relation to its environment, promoting a sense of ease and organisation in the thoughts of your mind. This dual effect not only soothes the mind's survival states but also cultivates the curious and receptive mental space needed to heal yourself or interrupt unhelpful thought patterns. Through repeated practice, your mind learns to associate these sensory and proprioceptive cues with a calm and open disposition, making the 'switch' from survival to ease more accessible over time. Without switching, your mind is focused on interpreting and making meaning of the feelings in the body and, as you now know, this isn't helpful. With *switching* it will be focused on receiving proprioceptive and sensory information. So, when you engage in these physical practices, instead of trying to make sense of the feeling of anxiety in the body, the mind will be focused on organising information coming from the body like what smell is in your nostrils, the sensation of touch on your skin or how it feels for your muscles and joints to be focused on balancing on one foot.

> **When you feel like you are stuck in your mind, switch into your body.**

Five years ago, Jen was working with patients in a neurological rehabilitation physiotherapy placement in a hospital. There was a young man, Tom, *who had suffered a traumatic brain injury. His initial symptoms were low mood and rage, both as a result of the brain injury and the accompanying loss of mobility and speech. After working with him on proprioceptive and vestibular-based exercises to improve balance and mobility, the clinical team noted that his mood would improve considerably. A simple round of balance and coordination exercises had an enormous impact. Not only would he be significantly more motivated to attempt mobility and independence, but Tom was also more calm and open to human interaction and connection. Over time, these exercises became an anchor in his recovery – allowing him to tap into the healing superpower his body already had: switching. As his vestibular and proprioceptive connections improved, his recovery moved at a much faster pace than expected and he was able to regain mobility, basic speech and coordination. Jen and Tom had a heart-to-heart one day and Tom shared some of the difficult family background and trauma he experienced as a child. He wasn't just healing from his injury, he was on a transformative journey, healing not only brain connections, but all that came before and the awareness of this active conscious healing would impact on him forever. Tom had an extremely successful brain injury recovery thanks to an interdisciplinary team, and the positive impact of the proprioceptive, balance and coordination exercises will always stay in Jen's memory.*

Although proprioceptive therapies are commonly implemented to improve motor function after an injury, we know that proprioceptive input to the brain from muscles and joints plays an important role in self-regulation and can evoke a positive shift in your mind and a vast range of emotions.[5] Intentional changes in movement can be as small as adjusting your posture, opening up your chest and taking large, deep breaths to increase confidence, or intentionally smiling to improve mood and mind state. In essence, making a switch uses the two-way relationship between the body and the mind, tapping into the body's inherent wisdom to guide the mind towards a state of tranquillity and exploration with minimal effort.

Your posture carries memories that your mind may have dissociated from or suppressed. Changing your posture changes the patterns you are stuck in.

Try it now:

1. Pause and notice your mood, the shape of your posture and where there is tension in your body.
2. Next, as if you were a cat waking from a nap, take a full 30 seconds to move and stretch your whole body: face, neck, shoulders, arms, hands, fingers, torso, hips, legs, feet and toes.
3. And now gently guide your body into a upright but relaxed posture.
4. Check back in with your mood, shape and tension.
5. How different do you feel now?

You're probably already more present, more calm and more aware. You successfully interrupted the state of your mind through your body – you're learning the language of your body. Good job!

Learning to switch can help move you from a disconnected mind-body conversation toward a connected, fluent one. Just as different languages convey unique meanings and nuances, these two aspects (your conscious thoughts and your experiencing body) communicate in their own ways, shaping your perceptions and overall well-being. The conscious awareness of your mind encompasses positive and negative thoughts, beliefs and your perceived and constructed sense of self. The conversation in your body, on the other hand, is related to your physiological and neural processes that communicate intrinsically and intuitively. The secret language of the body is to merge these two native languages that are part of you into one new dialect through which to converse in a fluent and harmonious manner.

The brain is constantly occupied in interpreting the experience of your body and the environment around it. As you know from the listening practices, this involves the transmission of an enormous amount of information at all times that you can tune into by placing your attention on BASE. Noticing BASE and exerting control over the state of your nervous system to begin to heal yourself can be challenging if you are trying to do so solely from your mind. Thanks in part to philosophical perspectives like those of Descartes, we've come to treat the mind as superior in the Western world, rewarding cognitive processes over emotional and physical awareness. Because society tends to view the mind as superior to the body and rewards us more for thinking than for feeling, our brains tend to form our 'identity' in thinking rather than feeling, in mind rather than body. If and when we experience intense negative emotions, they can become unbearable and our nervous system becomes desperate for a means to escape its own feelings. Whether through repression or dissociation, both seek to avoid this intense and unbearable experience within the body because it has become too overwhelming. But those feelings don't go anywhere, they are still there in the body - you just become cognitively unaware of them. This strategy is adaptive ... until it isn't.

To illustrate this, imagine a person named Jade, who represents the experience of thousands of our clients. Jade grew up in an environment where expressing emotions was discouraged. Whenever Jade felt sad, angry or anxious, her parents would dismiss her feelings and tell her to 'toughen up' or 'stop being so sensitive'. As a result, Jade's nervous system learned that feeling emotions and being vulnerable was not only inappropriate but also dangerous because it resulted in criticism from her parents and a deep sense of shame and rejection in her body. Jade then learned to repress these feelings and – as a means of fitting in with her family and receiving praise – developed the coping pattern of being a 'good girl' and overachieving. This was the best adaptive mechanism her adolescent nervous system could come up with.

As Jade entered adulthood, this pattern of emotional repression continued. For instance, when she failed to make the honour roll at her university, her body experienced strong emotions of sadness and shame but Jade didn't even feel them because she automatically suppressed them without knowing it, resulting in feelings of anger. Moreover, she coped by studying even harder and becoming more strict with her diet in an effort to be the smartest and look the best. This form of chronic repression and coping eventually led to negative consequences on her mental and emotional well-being. Her repressed feelings didn't just go away, they continued to dysregulate her nervous system.

Her coping patterns, though adaptive in the sense that they kept her functioning and achieving, were becoming maladaptive as they started to weaken her health. Eventually, she started experiencing physical symptoms including chronic migraines, IBS and back pain. In this example, the chronic maladaptive coping mechanism of emotional repression hindered Jade's ability to navigate life's challenges in a healthy and resilient way. By repressing and coping, Jade successfully distracted herself from facing the deep emotional pain in her body that was dysregulating her, but in avoiding that pain, she prevented herself from actually healing.

As you continue learning the language of your nervous system you will probably recognise in yourself many of its adaptive strategies. Hopefully, you will realise that some of the persistent states of survival, such as anxiety, depression and chronic illness, that you experience do not mean you are broken and stuck forever – they simply mean you are in need of a switch. Your nervous system is malleable, just like the plasticity of your brain. Under the right conditions and with practice, you will become the leader of your nervous system and help rewire it for persistent states of ease, comfort, flexibility and resilience. It will be like switching on a light.

Your Body Holds the Answers

'My mind needs help!' read the subject of Aaliyah*'s email. In early 2021 we met a wonderful woman in her early thirties. She was encouraged by another one of our clients to ask for guidance and help her with intrusive thoughts, anxiety and panic attacks. In her first message she explained that she had a history of chronic anxiety and that it was her goal to be able to go back to work after a six-month mental health leave. She explained that she had tried different therapeutic avenues over the years, but nothing had truly resolved her anxiety. In the very first session, Jen asked her, 'Would you be your own best friend if that best friend was the voice you speak to yourself every day?' She immediately said 'Yes!' and then after a few moments said: 'Maybe? I just thought about it, and I'm not sure what my inner voice is saying.' Aaliyah was in survival mode and her mind was running on default survival patterns that she wasn't yet aware of. We started coaching her into BASE-M and she learned that her inner voice was hypercritical, judgemental and saying things like 'Everything has to be perfect, or I'm a failure and*

nobody will love me', 'I'm worthless' and 'I'm nothing', giving her feelings of anxiety. She definitely did NOT want to be best friends with it. Once we helped her become aware of that mode of her mind, we began switching. Aaliyah was coached to do multiple gentle listening check-ins per day, and when she found that her mind was in an unhelpful mode she would switch. She used the proprioceptive exercises together with the touch ones as one single sequence and then she would use the five senses every morning and evening to get her brain in a helpful state. Aaliyah was extremely successful in interrupting the mode of her mind with the switching step of A I R. By stopping the dysregulating thoughts that were on a loop in her mind by switching into her body, she stopped her panic attacks and got back to work. It was then through the redesign step of distancing (which we will discuss in the next chapter), rewiring her default patterns through shifting the perception of herself and repairing the relationship to her younger self that she experienced long-term changes to her intrusive thoughts and healed herself. Aaliyah doesn't live in survival mode anymore. In fact here is a message she sent us recently:

'Hey you two! I cannot thank you enough for the gift you have given me of healing myself. I didn't know that all of THAT was already available inside me. It has changed every aspect of my life. I haven't had a panic attack in two years and if my anxious mode switches on, I know why I'm being triggered and I know what to do and how to soothe myself. My mind and body feel like home for the first time in forever. My inner voice truly has become my best friend! Thank you for changing my life, Aaliyah.'

Switching shifts you out of survival patterns, activates regulating physical pathways *and* calms the mind in such a way that it moves its focus from surviving to receiving sensory input from the body. So, you will not only be able to directly influence the state of your mind but, like Aaliyah, influence the stress, anxiety and unresolved patterns you've been trying to heal. Your mind will be redirected from being a constantly churning engine, always seeking meaning and interpretation in every situation, to more focused, present and in a grounded sensory-aware state. By literally taking action and moving out of your cognitive awareness and into your sensorium of awareness, you can redirect your energy from fuelling dysregulating processes to supporting regulating processes. In doing so, your brain gets to perceive clearly what's *actually* going on in your body and environment as opposed to being distracted, overwhelmed and caught up in the noise of your mind. This is of vital importance because most of the time, unless you truly are in a life-threatening situation, the reality is that you are physically safe. You are already at home in your body. But because your nervous system and mind are stuck in survival mode, they are incapable of recognising this, and can convince you that danger is imminent, even when it's not.

In our clinical experience, more often than not, the information that comes flooding into your brain through switching exercises is positive, which creates spontaneous regulation and primes your nervous system for healing and change. We go from living in a chronically looping alarming *story* of what's going on, to living moment to moment in the truth of *what's actually happening.* By tapping into the adaptability and malleability of your body and nervous system, you can find a pathway to clarity, resilience and well-being. So, when you feel overwhelmed, remember: your body holds the answers. Although it may seem daunting that the conscious mind you are familiar with is not the solution to initiate regulation and self-healing, we'd invite you to consider it as a hopeful revelation that engenders self-understanding, forgiveness and compassion. Most of us have struggled to heal ourselves by using our mental thoughts to wish/make/force ourselves to be/feel/

act in different ways to no avail. The failure and frustration of these efforts usually lead to even more negative thoughts in the form of guilt, shame, anger and self-blaming that reinforces the exhausting feedback loop of dysregulation. You may begin to understand why you've fallen short up until now. Realising that the mind is the most difficult place to originate healing should actually come as a relief! You didn't do anything wrong. You haven't failed. You are not broken. Switching from harried thoughts to a fluent, embodied conversation will help you begin to heal from whatever is keeping you stuck.

Let's summarise what we've discussed so far about Switching for Interruption:

- You may think you need your mind to change your thoughts, but in fact you need your body.
- The mind is a meaning-making machine and when it is unaware of the language of the body, it misinterprets its messages. This is how you get stuck in patterns like anxiety, stress and unresolved trauma.
- Switching taps into the regulating power of the somatosensory system through proprioception, touch and your five senses.
- Through the following exercises, your mind makes the 'switch' from survival mode to helpful mode and from dysregulating patterns to regulating ones.
- When your nervous system begins to access more helpful neural pathways, you begin to be fluent in the healing mind–body–human language.

Practices 2

Interruption: Switching

Learn to regulate your mind by switching into the body

In these practices you will learn three types of somatosensory exercises through proprioception, touch and your five senses. In each, you will be given various exercises that can either be done independently or as one long sequence.

Why Do It?

The switch techniques will help you interrupt unhelpful thought patterns in your mind by leveraging your somatosensory system. By repeating these, your mind is receiving direct information from the senses of your body, such as its position in space and the sensory feedback on your skin. This helps your mind organise information based on your physical reality rather than on the stressful loop of messages it's stuck in. This automatically tunes down the feelings of stress and tunes up feelings of safety.

When to Do It

When you are aware of unhelpful thoughts that are keeping you stuck in nervous system dysregulation, for example when you experience procrastination, anxiety or a state of freeze in your body - or when you're trying to listen but you're finding it difficult.

Tips before you begin

- When you notice your mind stuck in unhelpful patterns, use any of the switch practices to help support your nervous system regulation. These are very effective in helping you switch out of overthinking, over-analysing, ruminating, numbing, dissociating and catastrophising and into mental organisation mode and clarity.
- If you find yourself questioning specifics like level of pressure, repetitions per day, angle of arm position – know that this work is experiential and experimental. You are tailoring it to work best for you. Focus on achieving a *switch* in your mind and a shift in your nervous system, without over-complicating it.
- Your personal approach to these shifting practices may change over time and you may find yourself preferring one practice over another. Do what your mind and body find most helpful. As you continue to master the mind–body–human language of your nervous system you will find yourself aligning more and more with your inner compass, your intuition.
- Pay close attention to BASE before and after each exercise; you will find that even subtle actions can create noticeable shifts!

Practice 1: Push

1. Find a wall that you can press against.
2. Stand up straight.

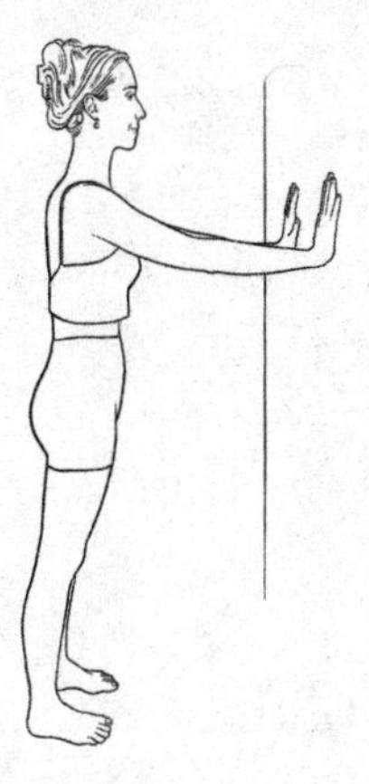

3. Push your hands up against a wall and let your whole body weight lean in as you bend your elbows to a degree that's comfortable for you.

4. Stay here for up to 10 seconds.
5. Rest for a moment.
6. Push your hands up against a wall again and lean in, and this time lift one knee towards your belly button and feel your core being engaged.

7. Repeat on the other side.

8. Do this exercise up to five times.
9. Notice how different you feel.

Note: A quick yet effective version of the push exercise is simply pressing your hands against each other for up to 30 seconds.

Practice 2: Thumb Circles

1. Extend your arms out in front of you with each hand in a thumbs-up position.

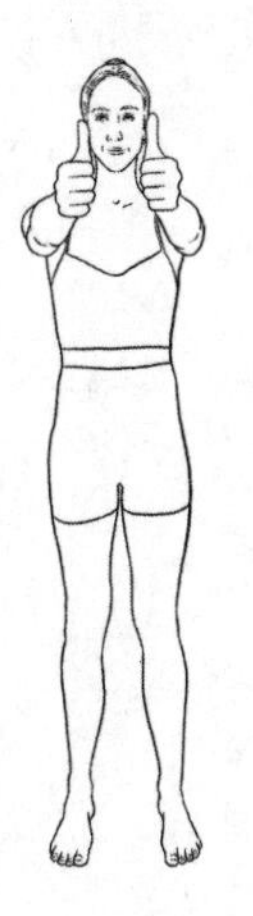

2. Then, draw a circle to the right side of your body with one thumb while your eyes (only) follow the movement of the thumb.

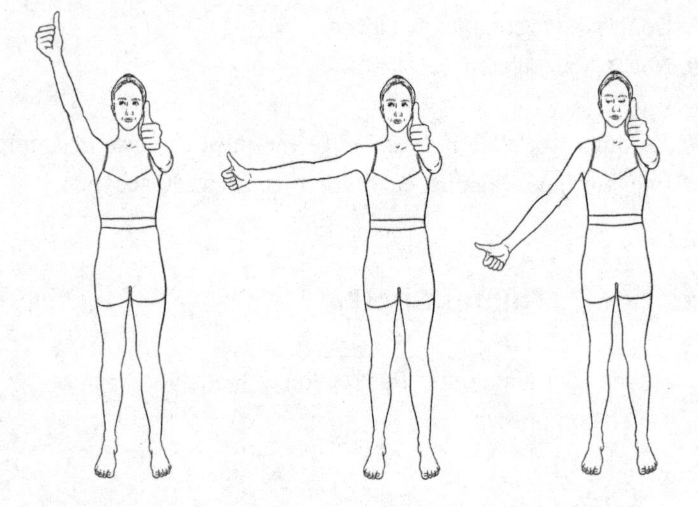

3. Repeat on the other side.

4. Do this three times.
5. Then, repeat the whole exercise while bending your legs in a squat position.

6. Notice how different you feel.

Practice 3: Fingers to Nose

1. Stand up.
2. Close your eyes.

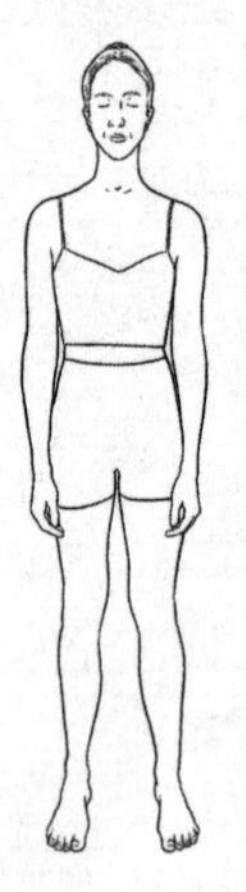

3. Then, touch your nose with one index finger while the other hand is extended out laterally.

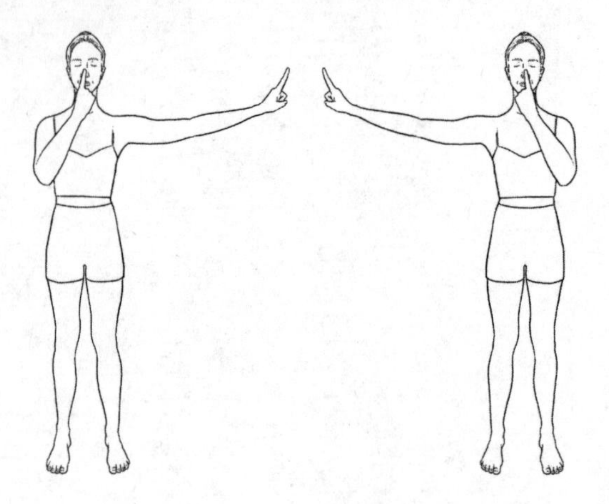

4. Repeat on the other side.
5. Do this alternating sides for up to 30 times.
6. Repeat the whole exercise while standing on one leg, and then the other leg.

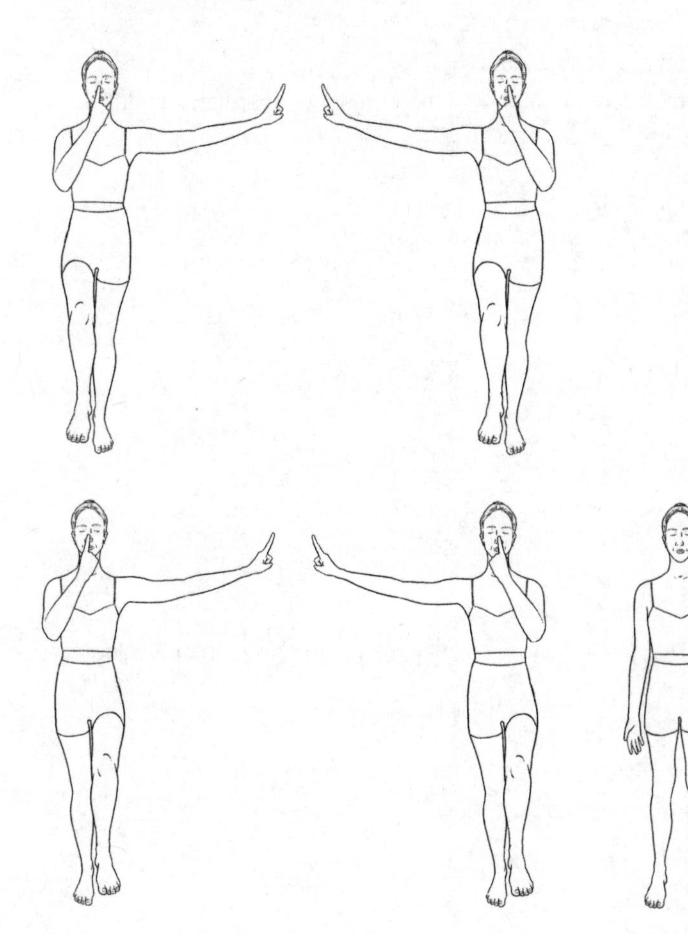

7. Notice how different you feel.

Practice 4: Touch

Feet

1. Stand up.
2. Take a deep breath.
3. Take a moment to feel the contact of your feet with the floor.

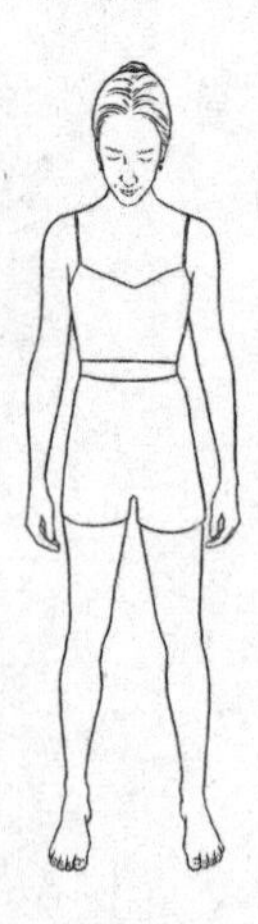

4. Then, sit down and gently massage one foot for 1 minute.

5. Stand up and notice the difference between your feet, your legs, your shoulders and even the way your ribcage feels on one side versus the other as you breathe.

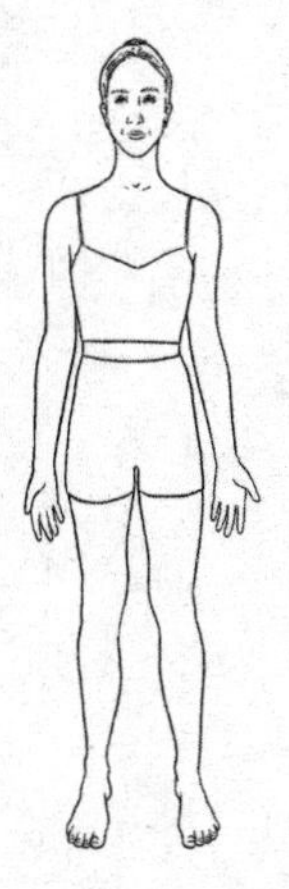

6. Sit down again and gently massage the other foot for 1 minute.
7. Stand up and notice how different the contact of your feet with the floor feels now, as well as your legs, your shoulders and how your ribcage feels as you breathe.

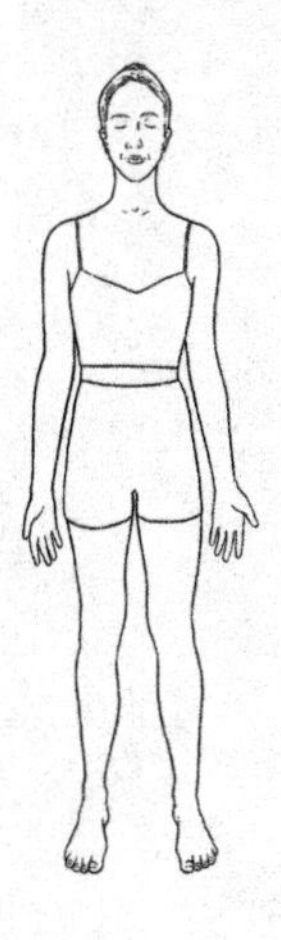

Arms

1. Sit down and allow your arms to hang by your sides.

2. Take a deep breath.
3. Take a moment to notice your arms and how they feel.

4. Then, begin to gently massage one arm for 1 minute.

5. Return to the hanging arm position and notice the difference between your arms, shoulders, neck and breath.

6. Gently massage the other arm for 1 minute.
7. Return to the hanging arm position and notice your arms, shoulders, neck and breath now.

Practice 5: Five Senses Experience

This exercise will guide you through an immersion into your five senses. As you proceed through each sense, do the following:

- Find a comfortable position to sit in.
- Notice how your body feels before you begin.
- Engage with the sense with lots of curiosity, as if you were noticing it for the first time.
- Do not judge what you are perceiving, just perceive it.
- Let your attention explore the 'detail cues' for each sense.
- Indulge in each sense and detail cues for 1–2 minutes.
- Notice how your body feels after each sensory experience.

1. Eyes and Sight
   Detail Cues: objects, emptiness, colours, shapes, light, shade, near, far, narrow, broad, what draws you, what repels you – can you feel what you see in your body?

2. Ears and Sound
   Detail Cues: loud, quiet, near, far, high, low, natural, mechanical – can you feel what you hear in your body?
3. Nose and Smell
   Detail Cues: odours, strong, faint, moisture, dryness, temperature, nostrils, sinuses, throat – can you feel what you smell in your body?
4. Tongue and Taste
   Detail Cues: tastes, flavours, strong, faint, moisture, dryness, temperature, mouth, throat, stomach – can you feel what you taste in your body?
5. Body, Touch and Felt Sense
   Detail Cues: air on skin, clothing on skin, surface contacts, softness, hardness, stiffness, pliability, temperature, movement of breath – can you feel what you touch in your body?

Make Switching a Part of Your Life

The goal of switching is to practise using the body to help regulate your mind when it can't regulate itself. To do this, in addition to the practices above, we invite you to explore a variety of somatosensory and movement practices. Some of our favourites include:

Proprioception

For cultivating proprioception, some of our favourite activities include yoga, dance, qigong and lifting weights. Regardless of what activity you choose, the most important element is that you derive fun and joy out of it.

Touch

For becoming more comfortable with touch, some of our favourites include self-massage (like the MELT Method, which is a neuro-signalling self-treatment clinically proven to reduce chronic pain and enhance well-being), therapeutic massage therapy and spending more time hugging and making contact with your loved ones. When it comes to touch, it's important that whatever you choose to do feels safe, makes you feel comforted and that you do it at a pace that works for you.

Five Senses

To become more engaged with your five senses and benefit from the spontaneous regulation they can provide, we suggest the following. Walking barefoot (especially in nature), listening to pleasant music, foraging, cooking and baking. Make sure that when doing any of these, you direct as much of your attention as possible to your sensory experience.

Notes:

....

....

....

....

....

....

Chapter 3: Redesign

Distancing

to observe yourself with compassion

To change ourselves effectively, we first have to change our perception.

Stephen R. Covey

In the last chapter, we learned how to interrupt the survival mode of the body by accessing regulation through our somatosensory system. This helped switch the mind from being in an unhelpful mode to a helpful mode where it could more accurately interpret the signals coming from your body through *listening*. In this chapter, we are going to learn the first Redesign technique in A I R called *distancing*. Building upon the more helpful and regulated mind cultivated through *switching*, *distancing* helps us get a better perspective on what is going on inside our mind and body by getting us 'out' of both our dysregulated thought processes and stress responses in the body so that we can more clearly see what's happening and make choices about how we want to help ourselves. Distancing does this by activating the part of us that is the best interpreter of the language of the body and the best agent for regulating your nervous system: your Observing Self.

The Observing Self

The Observing Self is our term for what is commonly referred to as 'witness consciousness'. Other names include the *true self*, *original self*, *atman*, *spirit* or simply the *self*. Every culture and spiritual practice has a different label for it but, regardless of the name, the Observing Self is pure awareness, the part of a human being that is capable of observing their own thoughts and embodied experiences. For a person to be able to conduct truly transformative healing work, whether that be in the spiritual, mental or physical realm, access to their Observing Self is fundamental. It is only through the Observing Self that we can step out of our dysregulated way of being, observe our default unconscious patterns and then reshape ourselves by providing different input to the whole system. Think about stepping outside yourself and watching your life and actions from an objective perspective. Difficult, right? But quite the super power if you can pull it off.

To put this into perspective, think about someone who's a workaholic. When we look at them from the outside, we can clearly see that they're overworking themselves, neglecting their personal life and heading for burnout. We might be in disbelief at how they're willingly sacrificing their well-being for their job. Yet, from their own perspective, they might be deeply immersed in their work, feeling the pressure to excel or succeed. They may not recognise the toll it's taking on their health and on their relationships. When we point out their excessive work hours or the negative impact on their life with care and concern, they might respond with a mix of reactions like denial, rationalisation and defensiveness. It's often only after they've reached a breaking point or experienced burnout that they can look back and see the situation for what it truly was, just as we could from an outside perspective. This is precisely the same dynamic that is going on inside us when we have a dysregulated nervous system. Since we are *in it*, we can't see clearly. And if we can't get outside the system and view things clearly, we can't possibly hope to change it. This is why the Observing Self will help you see things you wouldn't otherwise notice and, most importantly, it will help you step outside the patterns that keep you stressed, anxious and stuck.

Though many of us are familiar with the non-judgemental observer from mindfulness meditation, the Observing Self is much more than that. Stephen Cope, in his masterpiece *Yoga and the Quest for True Self*, describes the witness/observer as having the following qualities:

- **The witness does not choose for or against any aspect of reality.** The witness does not split life into good and bad, right and wrong, high and low, or spiritual and not-spiritual. The witness does not take sides, but experiences a kind of 'choiceless awareness' [though we would add that the witness can help cultivate and nurture useful vs less useful nervous system states].
- **The witness does not censor life.** The witness allows all thoughts, feelings and sensations to receive the light of awareness, without discriminating. There is absolutely nothing that the witness cannot see, feel and experience. There is no shadow, no shame, no repression that is not capable of being penetrated by witnessing. The witness is not judgemental in any way, but practises self-observation without judgement.
- **Witnessing is a whole-body experience.** Witnessing is not an intellectual exercise. Quite the contrary. We actually witness our experience with our whole body, allowing ourselves to feel the reverberations of sensations throughout the whole physical-emotional organism.

- **Witness consciousness is always present at least in its potential form in every human being at every moment.** The witness is the essence of our divine, awake, already enlightened nature. We don't have to create the witness. This quality of consciousness needs only to be recognised, evoked, claimed and cultivated.
- **The witness is the part of the already awake mind that is capable of standing completely still, even in the centre of the whirlwind of sensations, thoughts, feelings, fantasies – even in serious mental and physical illness.** From the witness, we can stand back and objectively observe our experience even as we're having that experience. Even as the witness stands as the still point at the centre of the storm, though, this part of our consciousness can fully 'dance' with life, directly experiencing all sensation, all activity in the mind, the heart, the body. It moves with experience, even as it remains completely still, anchored and grounded.
- **The witness goes everywhere.** The witness is connected to the whole quantum field of mind and matter. Witness consciousness stands outside time and space, living in the eternal now of the unmanifest realms, while also penetrating time and space. Witness consciousness is the quality of the self-aware universe. It is the intelligence, the 'sight without a seer', that saturates the whole quantum field of mind and matter.

Distancing is the technique we teach to empower you to become your own Observing Self and access all of the remarkable qualities that Stephen Cope describes. *Distancing* practices strengthen the brain's capacity for self-awareness and embodiment rather than thinking and distraction. Self-awareness is a fundamentally different mode of the brain than thinking. Self-awareness is associated with perspective, insight and regulating and soothing brain waves versus overthinking, languishing and the highly active brain waves generated by being in survival mode. Learning the skill of distancing will teach you that not only are you more than your thoughts, but you can rise above them, redirect them, and use them to help your nervous system influence your body's physiology and your sense of calm.

Mind, Brain and Spirit

We will be exploring the Observing Self from three lenses: mind, brain and spirit. We will first look at it from the psychological (mind) lens. Second is the neurological lens, as it is only in recent times that scientists have been able to peer into the brain and verify the physical proof of the impact of ancient witness consciousness techniques on our minds. Finally, we will look at it from the spiritual lens to see that it is from ancient wisdom traditions that we not only know about the Observing Self but also have practices for cultivating it.

First let's discuss the realm of the psychological (mind). Most forms of psychotherapy have a version of witness consciousness at their core. In fact, Sigmund Freud attributed aspects of witness consciousness to the ego, about which he said, 'the ego can take itself as an object, can treat itself as an object like other objects, can observe itself, criticize itself, and do heavens knows what with itself.'[1] At the heart of Freud's approach to healing was shedding light on the unconscious through free association and the observing ego. Half a century later in 1965, three psychologists, Arthur A. Miller, Kenneth S. Isaacs and Ernest A. Haggard, collaborated on an article that shifted the importance of psychologists' focus to working with clients on improving their awareness and functionality of their 'observing ego', which they argued is critical to our social functioning.[2] More recently, Richard Schwartz, the creator of Internal Family Systems Therapy and parts work, a remarkably effective method for healing trauma and becoming more 'whole', emphasises witness consciousnesses, which he calls *the self,* as the cornerstone of his method. He describes the self with what he calls the 8 Cs - compassion, calmness, curiosity, clarity, confidence, connectedness, creativity and courage - and the 5 Ps - patience, perspective, presence, playfulness and persistence.

A note on parts work: Parts work is a therapeutic approach rooted in the understanding that the human psyche is composed of a multitude of distinct yet interconnected 'parts'. These represent different facets of one's personality, emotions, beliefs and experiences. This work delves into the idea that these internal aspects can sometimes operate independently, each with its own perspective, desires and agenda. In this framework, parts are seen as sub-selves, each having its own unique role and function. Some parts may hold unprocessed emotions or past traumas, while others might embody protective mechanisms or coping strategies. These parts can influence thoughts, behaviours and emotional responses, sometimes leading to inner conflicts or patterns that affect a person's well-being and choices. The work of going from 'parts' to 'whole' involves a process of exploration, dialogue and integration aiming to understand and heal the underlying motivations. This process allows for a profound awareness of the intricate interplay between these parts and how they shape your experiences and perceptions, deeply influencing your nervous system. Parts work is a great self-awareness tool that we will be touching upon later in the book.

Another relevant psychotherapeutic approach based in witness consciousness and the Observing Self can be found in neuro-linguistic programming (NLP). At the beginning of the 1970s this innovative communication and therapeutic method became especially popular for its versatility in addressing a variety of mind and body illnesses, from anxiety and phobias to chronic pain and allergies. The co-founders, Richard Bandler, a linguist and John Grinder, a psychologist, recognised that there was an irrefutable connection between neurological processes (neuro-), language (linguistic) and acquired behavioural patterns (programming). NLP operates on the premise that individuals can rewire their thought patterns and behaviours by consciously adjusting the way they use language and their internal mental representations.

For instance, in a real-life scenario, if someone has a fear of public speaking (a common phobia), an NLP practitioner might help them identify the negative mental associations and language patterns contributing to their anxiety. Through techniques like reframing and anchoring, the individual can then transform their fear into confidence by changing their internal dialogue and mental imagery, ultimately enabling them to deliver a speech with newfound ease. When Jen was sick and bedridden, one of the first therapeutic approaches that actually worked for her was the Lightning Process, a therapy developed by psychologist and NLP practitioner Dr Phil Parker. One of the key transformative NLP components being used in this Lightning Process was something called perceptual positioning (PP). In 1987, this technique was developed by Grinder and Judith DeLoizer, a leading NLP practitioner. Judith's love of dance led her to integrate movement and the body as a primary tool in NLP. In PP, the subject physically moves around in space to different positions depending on what perspective they are adopting. Jen, a mover, dancer and somatic practitioner, immediately resonated with observing herself from a distance by literally moving her body and stepping into a different position in space and some of her approach to this work would be inspired by PP. Karden, whose background also includes movement-based therapies, also found this work to be influential, leading him to incorporate movement and somatics as a big part of his work.

Moving to the lens of the brain, neuroscientists have been using functional magnetic resonance imaging (fMRI) to measure how mindfulness practices impact on the brain and their findings have been astounding. When it comes to dysregulation vs regulation and surviving vs thriving states, there are two primary areas of the brain involved. On the survival side, the brain has the amygdala and hypothalamus - both of which are housed in an ancient part of our brain called the limbic system. They are constantly monitoring our environment for threats. The limbic system is hardwired to react if something threatens us, and it can do so stunningly fast. On the thriving side, in the neocortex, which evolved after the limbic system, is the medial prefrontal cortex (mPFC), referred to by some as the watchtower.[3] The mPFC observes the same stimuli that the amygdala does but it has much greater powers of discernment and the ability to inhibit and calm the 'shoot first, ask questions later' strategy of the amygdala.

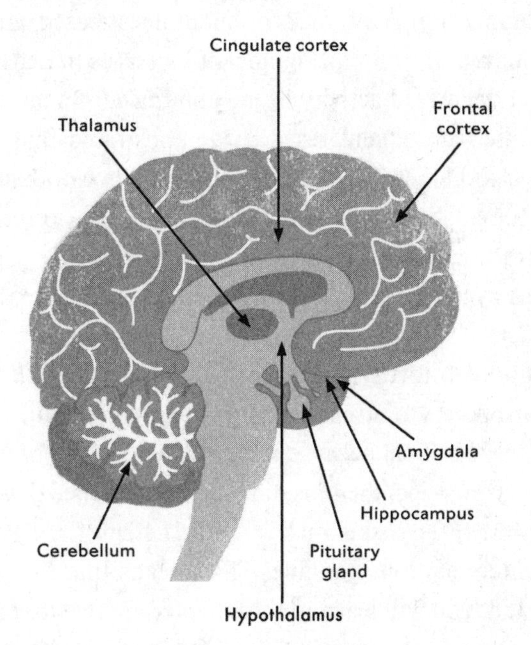

Functional MRI studies have shown that folks with dysregulated nervous systems tend to have overactive amygdalas due to chronic stress and trauma and much less active mPFCs. This is seen often in individuals with post-traumatic stress disorder, social anxiety or certain phobias.[4,5] That literally means that even if you are aware that you are having a dysregulated response to a particular situation, the part of your brain that is responsible for calming down your response *cannot do its job.* This is the reason your surviving mind is an unfavourable place to initiate self-regulation of the nervous system. However, if you fall into this category, don't worry, this is not your destiny. Studies have found that the skills you are learning in this book increase mPFC activity and decrease amygdala activity, as well as leading to a decrease in negative effects.[6]

For example, when neuroscientists studied what happens to people's brains after eight weeks of mindfulness-based stress reduction meditation, they found significant increases in activity in the mPFC and decreased activity in the amygdala.[7] On an even more hopeful note, subsequent research demonstrated that the same practices also helped the brains of individuals who had suffered severe trauma.[8] Strengthening the mPFC is like tuning up the brakes on the car of your nervous system, allowing you to slow down the amygdala, which is always trying to stomp on the gas for you to react.

In addition to fortifying the mPFC, Observing Self practices have been proven to strengthen a part of your brain called the insula.[9] This is critical because the insula is the area of your brain that allows you to feel the sensations and experience of your body. In other words, the insula makes the first step of A I R, *listening* through interoception, possible. That means that the *distancing* practices that you will learn in this chapter are two-for-one in that they strengthen both the mPFC and the insula at the same time! Activating the Observing Self with *distancing* practices and expanding embodied self-awareness with *listening* are the very heart of effective nervous system regulation.

In addition to the psychological and neurological foundations of the Observing Self, there is the lens of the spirit. As we mentioned, the idea of the Observing Self - *soul, spirit, atman* - is part of every culture and spiritual belief system. Across the world, the Observing Self has common characteristics that the Bhagavad Gita, perhaps the most famous Hindu religious-philosophical text, sums up as: 'The soul is never born nor dies ... Soul has not come into being, does not come into being, and will not come into being. Soul is unborn, eternal, ever-existing and primeval.' In modern parlance, it is the ghost in the machine, the ineffable awareness that is the *being* in human beings, the part of us that does the witnessing.

Practices designed to strengthen connection to this part of us such as meditation are recorded as far back as 1500 BCE, though it is safe to assume that they were being practised for thousands of years before the advent of writing. Of course, the most widespread and popular forms of witness consciousness practices are attributed to Buddhism. In the fifth century BCE as part of his Eightfold Path, the Buddha instructed his students in a practice called *bhavana* or 'mental cultivation' to aid 'the cleansing the mind of impurities and disturbances, such as ... hatred, ill-will ... worries, restlessness, skeptical doubts and cultivating such qualities as concentration, awareness, intelligence, will, energy ... confidence, joy and tranquility'.[10] Long before we understood the neurophysiology of the mind and body, the sages spoke the secret language of the body and invented nearly all the practices and frameworks, in one form or another, that we use for modern-day nervous system regulation.

> **Whether it be through the lens of the spirit, fMRI or psychotherapy, all agree that the Observing Self, the part of us that can witness the 'rest of us', is quintessential to healing our mind–body–human self.**

Let's Learn to Distance

A note from Sophia.

'Hi guys, I have been doing all the practices and stepping outside of myself so many times over the last few weeks. Suddenly something inside me clicked yesterday. I kept trying to fix myself and 'not think' what I was thinking and 'not feel' what I was feeling. And I realised I was pushing myself and trying to heal from survival mode. But yesterday as I stepped into Observing Self, I clearly saw that I was never supposed to be fixing myself. I was meant to learn what wound I have been covering up by burning myself out. Yesterday it dawned on me that ironically I found some love and validation inside me by stepping outside of myself. It was just there, hiding in a secret corner of my body and I felt myself begin to heal. Thank you to both of you for helping me fill myself up so I never feel empty again. You know this already ... but at first I thought recovery was going to be a combination of prescribed exercises and discipline. Recently it dawned on me that it is actually a deep getting to know myself. If you had told me this three years ago, I probably would have laughed. Thanks again for helping me meet myself so I can heal myself.'

Sophia is a registered nurse practitioner and lives her life fully empowered with the new-found knowledge that recovering from a chronic illness gave her. But this wasn't always the case. Sophia had found Jen and Karden on social media and had been following their content for a few months before reaching out. 'Jen, I am writing to you from my severe condition of being bed-bound and I see that you too were once in this state. I am a medical professional but never knew of any of these approaches you speak of. How do I learn what you did to recover?' Jen and Karden receive countless messages like this filled with questions and the outpouring of hearts and stories. But this one struck differently because she added, 'I have a masters in neurology and I have been a registered nurse practitioner for over 15 years. I never made any of the nervous system–chronic illness connections you share – and that I know of, neither have my colleagues.'

The lack of this knowledge among medical professionals is not an uncommon experience. Jen's first physiotherapy lecture at university wasn't about muscles or ligaments, it was about the biggest emerging public health issue: an ageing population and non-communicable diseases. It seemed like these lectures were going somewhere, but all they did was inform students of the fact that more and more younger people were living with and managing non-communicable stress-related chronic illnesses and more older people were living with multimorbidity. The lectures felt unproductive because all they did was illustrate the problem, like it was an unchangeable fact, as opposed to furnishing solutions. Jen remembers a debate that once happened. The floor was open for students to discuss some of the ways in which resolutions to this significant matter could take place. It was clear that the issue needed to be presented and discussed, but it was also clear that there was no existing textbook knowledge to turn to for solutions and thus nobody had any real answers. So, when Jen developed her chronic illnesses a few years into her studies, she didn't get angry at doctors who couldn't help her, she got angry at the system that accepted her condition as 'incurable' and had no actual understanding or solution to it. Sophia's message resonated with Jen in a deeply personal and professional way and so they began working together.

When she first started working with us, **Sophia** *was trapped in a cycle of exhaustion, pushing herself and crashing. Chronic fatigue syndrome had left her bedbound, with more than 30 symptoms, and it seemed like there was no way out. Doctors told her to keep* *resting and advised that with some rest she would recover. Sophia 'rested' for three years, and didn't recover at all until she found nervous system work. The approach we took with Sophia began with explaining the physiology of nervous system dysregulation. As with other doctors, psychologists, PhD neuroscientists, medical researchers and biologists that have come to us as clients we knew that the first important step for her was to deeply understand the science and logic behind the language of the body as well as review histories of other clients that were successful with our method. Then, once the belief and understanding is there, we move on to A I R. This story lives in this chapter because the Observing Self was what got Sophia from bedridden with fatigue to walking 5km per day in just a few weeks. We did this by coaching Sophia into learning about her beliefs and thoughts that were running on autopilot. 'I need to make sure everyone likes me. And I'll do this by helping everyone who asks me for help. I bet if I become a nurse, I'll feel complete.' Choosing nursing as a career coupled with a nervous system driven to help everyone around her to feel safe all the time was the perfect recipe for chronic stress and burnout. We coached Sophia on interrupting unhelpful thought patterns and to distance herself from them through the Observing Self. Each time she would occupy the mode of Observing Self and bring its qualities of connectedness, curiosity and compassion to mind and body, she found it soothing. In the beginning, she stepped outside herself up to 40 times per day, training her nervous system to become accustomed to a much more helpful mental dialogue and comfortable with states of ease. When she sent her note, we knew something profound was shifting for her and we were thrilled!*

Believe it or not, you have already been using a *distance*-like practice through the act of *listening* since it is inherently an act of witnessing and of paying attention to what you're noticing at any given moment in your body. That being said, *distancing* takes the witness consciousness to a new level. When doing this distancing technique, you are able to access the Observing Self mode and can gain an objective perspective of yourself. This is done via three fundamental components: space, contemplation and qualities of being.

Derived in part from perceptual positioning, space refers to the dimension or position that the Observing Self observes from. We live in a three-dimensional world and navigating this 3D world, as we covered in Chapter 2, is why we evolved an embodied nervous system. Therefore, incorporating concepts of space and movement in *distance* leverages primordial and powerful aspects of the brain's awareness to shift us from overthinking in survival mode towards the attentive and present mind of the Observing Self. So, unlike traditional witness consciousness practices that begin with silence and stillness, we will be teaching you how to incorporate physical movements in space to shift your body and mind into being a more helpful observer.

In addition to space, the contemplation component teaches you how to couple where your Observing Self is in physical space with a sensory-rich picture and feeling of the Observing Self. Generating this coupling harnesses the mind's qualities of attention *and* imagination, both of which help disrupt the cycle of negative thinking happening in the mind. Moreover, imbuing your Observing Self with a sensory-rich picture, which can include where it is (like a mountain top), how it looks (tall and graceful in flowing garments) and even how it smells (like a lavender field) activates more parts of your brain and nervous system, which tees up neuro- and bioplasticity, bringing your nervous system online into a more helpful mind and body experience. This part supercharges your access to the Observing Self and makes it more embodied and present.

Last but definitely not least of the components of *distancing* are 'qualities of being' imbued into the Observing Self. These are then broadcast back into the nervous system through its presence and observation. Though all the qualities that Richard Schwartz enumerated previously are important, the absolutely essential qualities of being are the three Cs of connectedness, compassion and curiosity.

Connectedness can seem paradoxical with a technique called *distance*, but it is only through stepping outside our patterns (distancing) that we can see them for what they are and heal them (redesign). Even though the Observing Self is positioned 'outside' us, as you read through Sophia's letter, it is not disassociated nor a bystander. It is our most loving self, our best friend, our biggest and most supportive fan. Its presence and 'with-us-ness' are unshakable, and it cannot be hijacked by your survival mode. It has the capacity to be with us like no one else ever could, it's just that most people are unaware of its presence and don't tap into it. The connectedness of the Observing Self can feel like the supportive presence of a friend that will always be there for you.

The second quality, compassion, is the most important quality. Without compassion, one cannot have true connectedness. Without compassion, we cannot truly *see* ourselves. Without compassion, the parts of us that are hurt, afraid and angry, and at the root of our dysregulation, will not express themselves. Compassion may be challenging to access for some of us because it was rarely given to us as children. Therefore, though we have always craved it, we don't know how to receive it or give it because it often wasn't modelled for us. The word 'compassion' originates from the Latin term *compati*, which quite literally means 'to suffer with'. This etymological root offers a useful insight into the nature of compassion as a process of sharing in the suffering or experiences of others. It highlights the idea that, just as we can empathise with the struggles and pain of those around us, we can also apply this same principle within ourselves. It's a way for the different parts of us to find empathy and compassion for other parts, especially those within us that might be distant or disconnected.

This approach reframes self-compassion as a gentle and healing process, one in which we guide and nurture ourselves through our own struggles. Instead of forcefully demanding love and understanding from ourselves, we're simply holding our own hand, providing support, and allowing for self-compassion to naturally emerge. In this way, self-compassion becomes a tool for self-healing, connecting the various aspects of our internal world, and fostering a sense of kindness towards ourselves. Swami Kripalu, a great yogic sage, believed that yoga students could not make progress in the development of witness consciousness without first intentionally cultivating self-compassion and softening the voice of their inner critic.[11]

The final essential pillar of the Observing Self is curiosity. As we mentioned in Chapter 1, when you are curious, you are primed to be in a regulating state for your nervous system. Curiosity pivots the mind from asking why (judgement) to asking what (discovery). Therefore, curiosity is a powerful complement to the acceptance component of compassion and supports non-judgemental observation. Furthermore, curiosity activates the brain centres associated with deep learning and memory which leads to the ability to make new choices and take new actions.[12] And when we are curious about something, it means that we are excited to engage with whatever we are about to observe, which is very helpful and regulating for the nervous system. The Observing Self qualities' that have been championed by spiritual sages for thousands of years align closely with the critical concept of co-regulation that we briefly discussed at the end of Chapter 1. Studies by modern psychologists and neuroscientists researching childhood development concluded that healthy childhood development was entirely dependent on the quality of the emotional attunement and attachment between children and their primary caregiver (typically their mother). We will go into this in much more depth in Part 3, but for now we'll simply say that safety and regulation in the nervous system is fostered by the attention (connectedness) and emotional responsiveness (kindness, care and acceptance) of the child's mother. When the Observing Self is imbued with these qualities, it is literally harnessing the essential ingredients of co-regulation that parents use to raise healthy, well-regulated children to allow your adult self to heal and regulate your nervous system in the present moment.

Let's summarise what you've learned so far about Distancing to Redesign:

- To truly heal and rewire the mind's survival mode, access to the Observing Self is fundamental.
- Distance is a technique that allows you to step out of your unhelpful unconscious patterns and objectively access your Observing Self, strengthening your capacity for healing.
- The Observing Self can be considered through the lenses of the spirit, brain and mind.
- Distance components include space, contemplation and the Observing Self qualities of being of connectedness, compassion and curiosity.
- The three Cs of connectedness, compassion and curiosity have been touted by spiritual leaders for millennia, while more recently have also been identified as keys to healthy childhood development. The methods of raising healthy, well-regulated children can also be used to allow your adult self to heal.

Practices 3

Redesign: Distancing

Redesign your mind by getting distance from its unhelpful patterns

In these practices you will learn how to redesign the patterns in your mind by helping it adopt an *observer perspective* and by learning how to cultivate your Observing Self.

Why Do It?

The distance techniques will help you redesign unhelpful mind patterns through exercises that are designed to broaden your perspective and perception of self. This new perspective is directly aimed at using the power of your mind to help you regulate your nervous system.

When to Do It

Any time you are noticing dysregulation in your body like excessive unexplained fatigue, unexplained gut issues, tension headaches, low back pain or unhelpful thoughts and behaviours like engaging in procrastination or people-pleasing patterns. If you encounter resistance in your Distancing practices, use the Switch practices first.

Tips before you begin

In these exercises, your mind may be tempted to overthink and over-analyse. This is normal. To help it stay on track, you can interrupt it anytime it veers into an unhelpful mode through placing your attention on your body with any of the switching exercises. The Observing Self can become the most useful mode of your mind because it is able to observe and guide you from a different perspective than the one that keeps you stuck. Practising these exercises will enable the neuroplastic changes in the brain allowing this process to become easier and more therapeutic over time.

Practice 1: Stepping Outside Yourself

1. Stand up and take a deep breath.
2. Take a moment to notice yourself through BASE-M.
3. Then step into a different position than the one you're standing in, directly opposite where you were before. Now you are looking at where you were standing as your Observing Self.

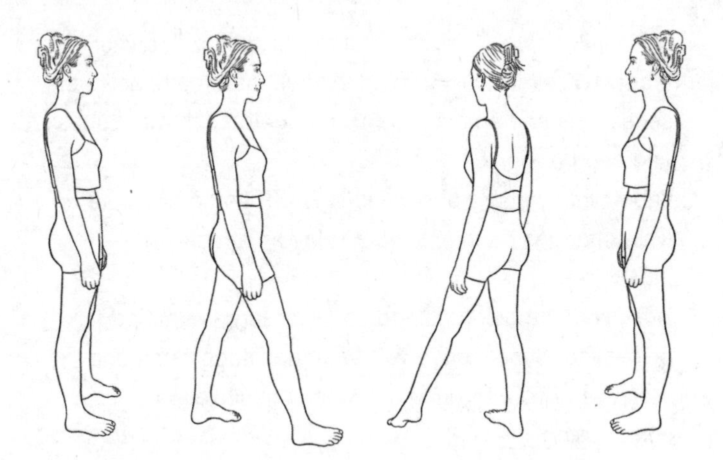

4. From this new place you are the observer, not the experiencer:
    a. Describe what you see of yourself in the third person as you observe and notice without judgement. E.g. 'I see that she is unhappy, she looks sad. I can see that her shoulders are rounded and that she doesn't want to stand up straight. I can see that she doesn't see herself as I do.'
    b. Take a few moments to observe yourself from here with curiosity, care and compassion.
    c. Notice how this feels in your BASE-M.

Practice 2: Cultivating Your Observing Self

1. Find a comfortable place to sit.
2. Use your imagination for 2 minutes to visualise yourself as an all-knowing, wise, curious, patient and compassionate sage on the highest mountain top capable of seeing everything.

3. From this perspective, imagine being able to see your triggered self far below you. Observe the breath, actions, sensations and emotions of this triggered self with curiosity and compassion.
4. Label what you are observing in BASE and make note of what changes happen in your body as a result of observing.
5. Next, imagine being able to see your triggered mind far below you. Simply observe the thoughts, opinions and narrative of your thinking self with curiosity and compassion.
6. Label what you are observing in your mind and make note of what changes happen in your body as a result of observing.
7. Enjoy this embodied experience of cultivating your Observing Self for another 1–3 minutes.
8. Smile.
9. Acknowledge yourself and appreciate yourself for taking the time to observe, connect and bring compassion to yourself.

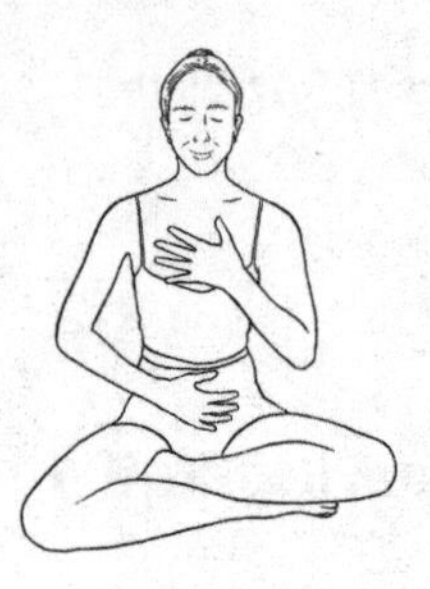

Make Distancing a Part of Your Life

The goal is to make the observing capacity of your mind your default mode. To do this, we invite you to:

1. Make the above practices part of your daily routine to strengthen your mind's capacity for observation and co-regulation.
2. Make a habit of stepping into your Observing Self whenever you are triggered.
3. Commit yourself to being the observer of your experience instead of the prisoner of your experience.
4. Spend more time using your mind to observe rather than overthink.

Notes:

....

....

....

....

....

....

Part II
Body

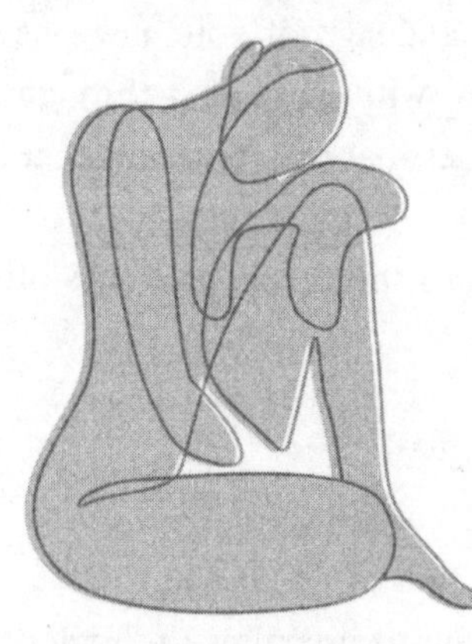

The body speaks louder than words.

Unknown

It is through symptoms that your body speaks its unspoken truths, and it is through understanding these messages that you will begin healing yourself and transforming symptoms into wisdom.

Like the mind, the body is a realm of memory, consciousness and identity, and has its own perceptions of the experience of your life and self - because you have a nervous system. When you are anxious and stressed, your nervous system communicates for you. It's deciding the experience you are having based on the survival responses it has learned to employ to keep you safe and alive. You will learn to speak the language of the nervous system and influence its responses to help you move from surviving to thriving.

You now know how to use the mind to listen to the messages coming from the body, transmitted through breath, actions, sensations and emotions. And like all conversations, simply hearing the words being said does not mean you are truly understanding their meaning. Underlying the messages is important information communicated in the non-verbal dialect that has been hardwired into us over millions of years of evolution.

You will learn how to decode the language of your nervous system and how to speak back to it, not with words but through direct physical communication techniques that help create change and calm your nervous system.

In this part of the book, you will learn the A I R practices of your body.

Translating

An **awareness** technique that teaches you to interpret the underlying survival logic of the language of the body.

Modifying

An **interruption** technique that teaches you how to change the survival mode your nervous system is in by physically and functionally stimulating your powerfully healing vagus nerve.

Settling

A **redesign** technique that teaches your nervous system how to make the feeling of safety, ease and embodiment its preferred state, helping you arrive home into your body.

Chapter 4: Awareness

Translating

the messages coming from your body

If you listen to your body when it whispers, you won't have to hear it scream.

Cherokee proverb

Now that you know the A I R techniques to begin regulating your nervous system from your mind, it's time to dive into the nervous system regulation of your body. We started by helping you understand and become aware that there is a conversation going on between your nervous system and your mind and how best to listen to and influence it. Now, in Part 2, we begin the transformative work of tuning into and changing the dysregulated conversation between your nervous system and your body so that you can bring harmony to your life. To tune into this conversation effectively, you need to understand what it's saying. *Translating* is the second **Awareness** technique that we are going to teach you so that you can identify messages in the form of sensations and feelings, and translate these into the survival logic of the nervous system, giving you the information you need to know what is actually happening inside you - and become fluent in the language of your body. To help give you a deeper understanding of what's actually going on inside you through your nervous system, we will examine the polyvagal theory. You will learn why your nervous system is doing what it's doing and, most importantly, what you can do about it to heal yourself.

As we covered in Chapter 1, your nervous system evolved for one reason and one reason only: to keep you alive. Following this prime directive, every structure and function of the nervous system is governed by the fundamental binary question of safe vs unsafe. Chapters 1 and 2 reviewed the evolutionary origins of our embodied autonomic nervous system, carrying out this safe vs unsafe directive for primordial species millions of years before humans. Over time, humans developed a cognitive mind capable of reflecting upon and interpreting this visceral, non-verbal process of your nervous system. But we have moved farther and farther away from understanding the embodied language of ourselves. In order to speak the secret language of the body you need to learn how to recognise and translate this ancient safe vs unsafe survival logic with the least amount of distortion by your cognitive interpretation. This is why we take you into the awareness of your mind before going into the awareness of your body. Without an aware mind, the work of regulation and self-healing is like trying to hear a whisper in a noisy room while you are listening to music on your headphones.

Dr Stephen Porges, the founder of the polyvagal theory (more on this shortly), and the Polyvagal Institute coined a term called neuroception. This term defines the survival-driven, non-verbal, automatic and instantaneous conversation and actions being carried out by your brain and body beneath your typical cognitive perception. To think of this in a real-world context, imagine we are on a hike. As we are hiking along a trail, one of your five senses, in this instance your eyes, notices a dark sinuous shape on the ground three feet in front of you. *Long* before your cognitive awareness registers this, your visual cortex has sent a message to the hypothalamus and amygdala. These two primitive brain centres, with their 'shoot first and ask questions later' survival strategy, determine that the sinuous line is a snake. In the meantime, your medial prefrontal cortex and cognitive brain structure are still lagging well behind. They are not in on this neuroceptive conversation as the autonomic nervous system initiates a recoil response through your neuromotor system, possibly causing you to scream and leap in the air to avoid the snake! A few milliseconds after your neuroception has taken the necessary action to save your life, your mPFC and rational brain functions catch up and notice that the 'snake' was in fact just a branch, there was nothing to fear, and over the next few minutes directs your nervous system to calm down and return to a regulated state.

The survival-logic-governed neuroceptive conversation in this hiking example is actually happening all the time - not only in acute circumstances. By *translating* what you are *listening* to in your body, you can learn what survival strategy (adaptation response) your ANS is deploying, which can be the cause of your dysregulation at any given moment (or extended period of time in the case of chronic conditions). Empowered with knowing what and why the ANS is doing what it's doing, you can then interrupt it with the upcoming *modifying* and *settling* techniques in Chapters 5 and 6. The key to decoding the survival logic, functions and nervous system presentations happening in human beings is best explained through the polyvagal theory.

Polyvagal Theory

Despite being the highest life form on the planet we still contain within us the genetic information and survival strategies of the animals that we evolved from. Evolutionary science has demonstrated that there are some shared survival strategies across all mammals, such as the fight or flight response, or a more dramatic freeze state.[1] While humans have evolved to develop more flexible and advanced cognitive survival strategies in addition, these fixed neural circuits are still hardwired in us. A beautiful and vivid testimony to this is to look at the developing foetuses of humans alongside our lesser evolved cousins like fish and reptiles. As you can see from the image, human foetuses look like the embryos we have evolved from.

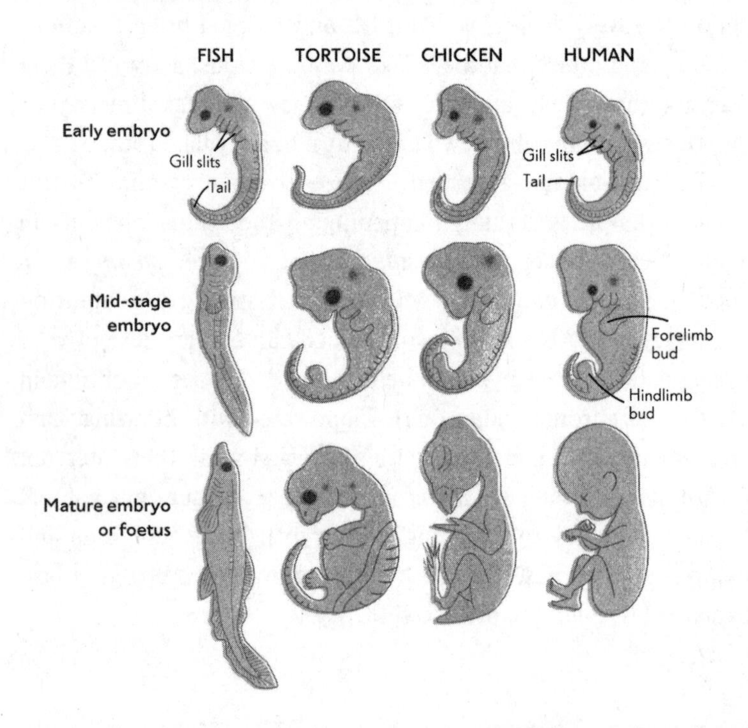

Most of your nervous system is an animal first and a thinking human being second. This explains why you can become stuck and sick without knowing logically why or how; the nervous system is responding for you before your mind has a chance to bring logic to a situation. The study of the evolutionary relationships among various biological species based upon similarities and differences in their physical or genetic characteristics is called phylogeny. In 1994, the aforementioned Dr Stephen Porges, a psychologist, neuroscientist, professor and researcher with more than 400 published scientific peer-reviewed papers, developed the polyvagal theory out of his experiments with the vagus nerve. The polyvagal theory proposes that the way our nervous system actually executes survival in our bodies is through a blend of three distinct neural structures and associated strategies that we have inherited and maintained (with some modification) from our ancestors as far back as when we swam in the sea. Dr Porges' breakthrough identified these neurological structures anatomically, showing their origins in our predecessor species and most importantly demonstrating how they actually govern human biology, physiology and behaviour under various levels of threat and play a big role in the mechanisms that underlie the relationship between chronic illness and trauma. The exploration of this theory has led to innovative treatments that target the mechanisms underlying symptoms in a range of behavioural, psychiatric and physical disorders.

The reason why polyvagal theory is called 'polyvagal' is because it is based on the anatomy of the vagus nerve. While the vagus nerve will be explained in more detail in Chapter 5, for now we want you to know the vagus is the tenth cranial nerve; it originates directly from the brain (not via the spinal cord), in the medulla oblongata, nestled at the base of the brainstem. Its name, vagus, translates to *wandering* in Latin, aptly describing its extensive network and branches throughout the body. Although we commonly refer to it as a singular entity it takes on the appearance of a dual, meandering cable, with one on each side of the body.

The vagus nerve encompasses the parasympathetic nervous system and divides itself into two main branches – the ventral (front of the body) and the dorsal (back of the body). The ventral

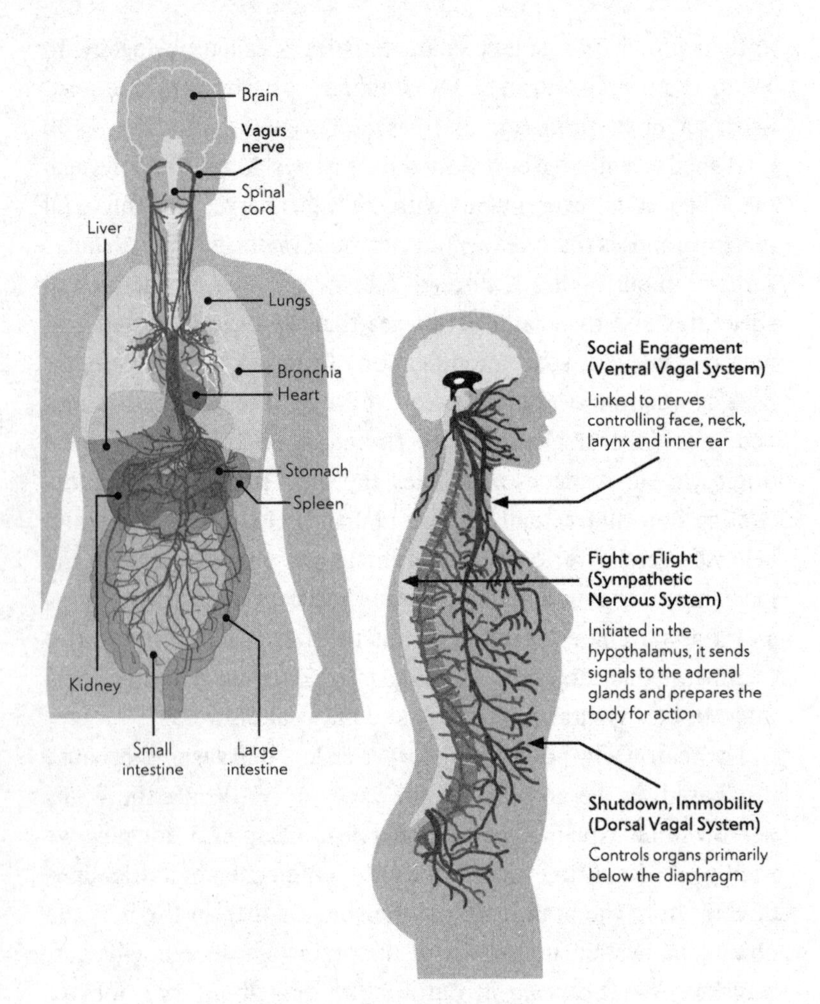

extends down through the neck, chest and abdomen, influencing and innervating multiple organs and bodily functions. The dorsal extends down the spinal cord and primarily influences and innervates the organs in the abdomen. The sympathetic response, on the other hand, is carried out by the HPA Axis and nerve fibres that emerge from the mid and lower spinal cord that project throughout the body innervating various organs and tissues. This anatomy is directly correlated to where the responses occur in the body.

With this basic understanding of the vagus nerve, we can explore the polyvagal theory, which says that when exposed to a threat, your nervous system will deploy a three-tiered hierarchy of responses to keep you safe, matched to the three sub-systems introduced previously. These responses are called the ventral vagal, the sympathetic and the dorsal vagal. The hierarchy of response is built into our ANS and anchored in the evolutionary development of our species. The origin of the dorsal vagal pathway of the parasympathetic branch and its immobilisation response comes from our ancient vertebrate ancestors and is the oldest pathway. The sympathetic branch and its pattern of mobilisation was next to develop. The most recent one, the ventral vagal pathway of the parasympathetic branch, enables the patterns of social engagement and connection that are unique to mammals.

At the top of the hierarchy, the parasympathetic *ventral vagal state* is present when you are *not* under threat. In fact, this state is also called the social engagement system, which inhibits our survival nervous system responses so that we can engage with and cooperate with our fellow human beings through facial expressions, vocalisations and coordinated movement. Without this system, the complex social societies of mammals, primates and human beings wouldn't be possible. When we are 'in ventral', we feel calm and comfortable, curious, connected to others, mindful and present. Our body feels relaxed, restored and able to comfortably engage in any social, physical or psychological activity. We are *regulated* and our mind and body are operating in a healthful nervous system state with optimal functioning of the heart, lungs and digestive organs.

Note on state: When we mention 'state' in the context of the polyvagal theory, we're referring to the physiological state of your autonomic nervous system. For example when you are in the sympathetic state your body is prepared for action. Your heart rate increases, breathing becomes faster, your muscles tense, and your body releases stress hormones like adrenaline and cortisol. When you are in the dorsal state, your body feels numb and unresponsive, your digestion, heart and breathing rate slow down and you feel socially withdrawn.

Our body also employs this mechanism to help others in achieving a sense of safety and connection, known as 'co-regulation', which we discussed in Chapter 3. Co-regulation allows us to help one another establish a feeling of safety and connection, achieved through actions like maintaining soft eye contact, tilting our heads as we speak, smiling, comforting a baby with gentle sounds and songs, and providing nourishment through breastfeeding. When you experience stress, the ventral vagal system conducts the first-level response, called 'tend and befriend', which is where we seek this social engagement. We look for connection with people, congregate in groups and take care of one another to mobilise to safety.

As your neuroceptive perception of threat increases, your ANS will switch off the ventral vagal state and activate the *sympathetic response*, which is commonly called the flight or fight response or the mobilisation response. At this point your intrinsic survival hierarchy has decided that the threat cannot be resolved by seeking connection, instead it needs to be resolved by running away and if running away isn't available or isn't the safest approach to get you to safety it will want to fight back. Your nervous system makes this decision based on your past programming (how you survived in the past is how your nervous system will keep you safe in the present). That being said, the nervous system and its mechanics are not black and white nor is the sympathetic response good or bad. In the flight response, the emotions can range from anxiety to terror and its actions can range from intense worrying to running for your life. The fight response has a range of emotions with frustration on the low end and rage on the high end, and actions can range from firmly establishing a boundary by speaking up for yourself to getting into a physical altercation. Unlike the ventral vagal state where your mind and body are running at optimal, the sympathetic response increases muscle tension, blood pressure, heart rate and blood clotting, and decreases activity in your digestive system and immune response. This is not a problem if this response is utilised as it was designed, for extremely short-lived, high-intensity moments of survival (followed by a return to the ventral state), but in a dysregulated nervous system, elements of your body's sympathetic response can become perpetual. On the outside, someone in perpetual flight mode may be stuck in avoidance and procrastination patterns, experiencing constant worry about most aspects of their life such as finances, relationships, health and work. Someone in chronic fight mode may be stuck in impulsivity, aggression and hyperactivity patterns. They may be excessively competitive and always pushing for perfectionism and achievement. On the inside, both of these patterns are felt in energy rushes in the arms and legs, heart palpitations, shallow breathing, a feeling of muscle tension in the back and neck, an increased sensitivity for pain and fatigue, and increased sweating and digestive issues. These are the body's whispers we will be *translating* in this chapter to help us prevent and reverse what is a perfect storm leading to burnout, pain and fatigue.

When this sympathetic state is activated chronically, you may see associations with a host of diseases including hypertension, atherosclerosis, arrhythmias, atrial fibrillation, insulin resistance, chronic inflammation and metabolic syndromes like obesity and high blood pressure. Aside from the usual solutions you've heard such as stress reduction and regular exercise, nervous system regulation is one of the most powerful ways to mitigate the adverse effects of chronic sympathetic states because it addresses the dysregulation at the root: the nervous system and the default patterns it uses to keep you safe.

Finally, moving down the hierarchy, if neither fighting nor fleeing is possible or successful, the sympathetic arousal can get so extreme that it is too much for the body to handle. At this point, we have a failsafe survival mechanism. The parasympathetic *dorsal vagal state* spikes and comes in so strongly that it overwhelms the sympathetic arousal. This is our most ancient survival strategy inherited from our cold-blooded ancestors - which is most commonly called the shutdown response. The most extreme manifestation of this strategy is fainting. When someone faints in response to a needle in the doctor's surgery, that is their neuroception perceiving the situation as overwhelmingly dangerous and instantly commands the body to literally shut down. Other ways we experience dorsal vagal state in the body is through a range of experiences in the mind and body and can feel like dissociation, helplessness, heaviness, numbness, passivity, lack of will, depression, hopelessness and despair. When it is chronically activated it decreases metabolism and gut function (often but not always leading to weight gain), and can also decrease heart rate, blood pressure, muscle tone, temperature and immune response.

But, like everything in the nervous system, things don't always happen in a predicted linear fashion. For example, most people who are victims of sexual assault don't faint, they usually freeze.

They report the inability (even if they cognitively wanted to) to fight back against their attacker. Freeze occurs when the body experiences the fight or flight response and the dorsal vagal response of immobilisation at the same time. Someone who experiences an event like this will likely have a nervous system that defaults to the freeze response in the face of stress, threats or triggers. This is the very essence of trauma: being exposed to a threatening event where none of the nervous systems' survival methods are successful, and as a result they get stuck and repeat the unsuccessful pattern anytime they feel threatened. Fortunately, forms of therapy that involve the body and autonomic nervous system, like somatic experiencing, help the nervous system complete these responses from the past while in the present, which can help to move them through and out of the body. This is how trauma can finally be healed.

The nervous system can get stuck in deploying the chronic freeze response in other ways too. Sometimes this looks like ongoing high levels of stress and anxiety, long-term grief or loss, or complex trauma leading to post-traumatic stress disorder. If you were driving a car, the freeze response would feel like pressing down the gas and brake pedal at the same time and you would be simply stuck yet exerting an enormous amount of energy while unable to move forward. This means that while the body is preparing to flee or fight, it's also stuck and frozen. This simultaneous build-up of energy for action coupled with the demand for immobilisation causes the heightened energy to become trapped in the nervous system.

While the polyvagal hierarchy of responses is a very useful model, they may not always present one at a time. Nervous system dysregulation can manifest as a blend of states responding out of sync. It's completely normal to be presenting with some sympathetic and dorsal vagal responses at the same time or in different

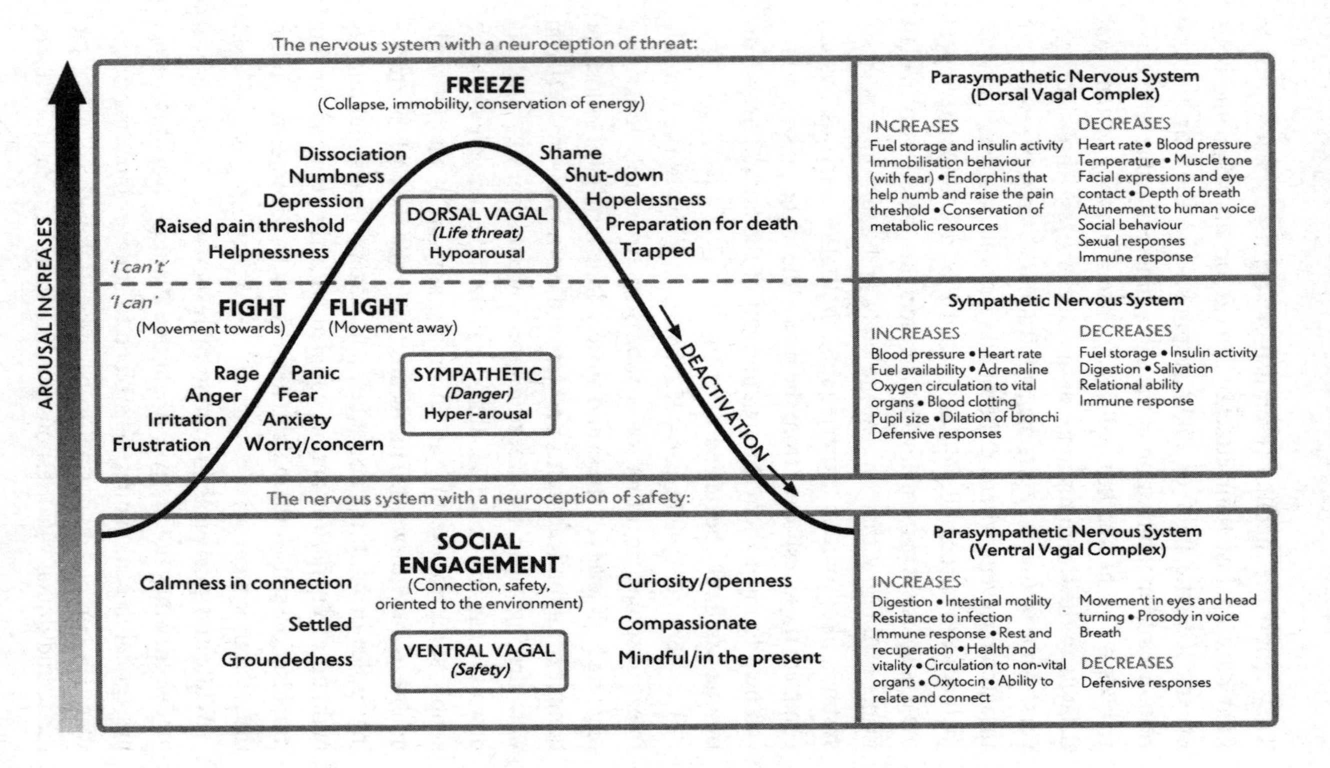
The nervous system with a neuroception of threat:
FREEZE
(Collapse, immobility, conservation of energy)
Dissociation
Numbness
Depression
Raised pain threshold
Helpnessness
Shame
Shut-down
Hopelessness
Preparation for death
Trapped
DORSAL VAGAL
(Life threat)
Hypoarousal
'I can't'
'I can'
FIGHT
(Movement towards)
FLIGHT
(Movement away)
Rage
Anger
Irritation
Frustration
Panic
Fear
Anxiety
Worry/concern
SYMPATHETIC
(Danger)
Hyper-arousal
DEACTIVATION
AROUSAL INCREASES
The nervous system with a neuroception of safety:
SOCIAL ENGAGEMENT
(Connection, safety, oriented to the environment)
Calmness in connection
Settled
Groundedness
Curiosity/openness
Compassionate
Mindful/in the present
VENTRAL VAGAL
(Safety)
Parasympathetic Nervous System
(Dorsal Vagal Complex)
INCREASES
Fuel storage and insulin activity
Immobilisation behaviour (with fear) • Endorphins that help numb and raise the pain threshold • Conservation of metabolic resources
DECREASES
Heart rate • Blood pressure
Temperature • Muscle tone
Facial expressions and eye contact • Depth of breath
Attunement to human voice
Social behaviour
Sexual responses
Immune response
Sympathetic Nervous System
INCREASES
Blood pressure • Heart rate
Fuel availability • Adrenaline
Oxygen circulation to vital organs • Blood clotting
Pupil size • Dilation of bronchi
Defensive responses
DECREASES
Fuel storage • Insulin activity
Digestion • Salivation
Relational ability
Immune response
Parasympathetic Nervous System
(Ventral Vagal Complex)
INCREASES
Digestion • Intestinal motility
Resistance to infection
Immune response • Rest and recuperation • Health and vitality • Circulation to non-vital organs • Oxytocin • Ability to relate and connect
Movement in eyes and head turning • Prosody in voice
Breath
DECREASES
Defensive responses

ratios depending on the triggering circumstances. For example, when facing a social situation that triggers anxiety, a person may experience a blend of sympathetic responses, such as increased heart rate, sweating and alertness. Simultaneously, they may also have dorsal vagal elements, manifesting as feelings of withdrawal, fear or emotional shutdown. The ratio of sympathetic to dorsal vagal response can vary based on the specific social context and the individual's prior experiences. In chapters 4, 5 and 6 of this book, we will be directing our attention at BASE-B, where B represents the polyvagal-based survival strategy being communicated by your body, and *translating* is the Awareness technique used to determine it.

The Body Whispers Before It Screams

Neuroception tells you that your autonomic nervous system will always respond before your mind has a chance to analyse, think and respond. When we don't speak the language of the body, the conscious mind can't understand - much less have an awareness - of what the ANS is doing. These might become out of sync and misunderstand one another. We may get distressing signals of anxiety and experience chest pain and tightness and think we are having a heart attack. But in reality our nervous system recognised a threatening pattern which initiated a cascade of stress hormone responses mimicking the symptoms of a heart attack. Really, your body is just trying to communicate an important message: 'I don't feel safe, and something isn't right.'

Safety in the body is the foundation of any positive experience and emotion. Without safety there is no joy. Without safety there is no confidence. Without safety there is no calm. Although life isn't all about positive experiences, we do need them to help balance out and cope with the negative ones. Nervous system regulation is about flexibility and balance. It's about experiencing stress and deploying a useful sympathetic response and then having a body that knows how to come back to calm. This is why learning how to interpret and translate the messages coming from the body will help you heal yourself and recognise that the sensations and feelings in the body mean something more than just *anxiety*, *depression* or *pain*. They may actually mean hurt, abandonment and not being 'enough'.

Jen's Diary

Journal entry: 22 June 2020

I lay there in the dark for what felt like hours, days ... maybe infinity. Whose was this body? It wasn't mine. Mine was made for dancing. It was made to kiss the earth and curtsey to the moon. That's what I was made for. I began drifting away and imagined myself on stage. I could feel myself move with such ease and grace that I melted into the air and turned into music. My mind drifted off, and my thoughts began taking on many different forms, like in that movie Fantasia *where elephants become flamingos and humpback whales soar into the sky. I let myself drown and sink to the bottom of an ocean. And as I became one with darkness, here on the ocean's floor, I let myself disappear. Just as my mind became quiet and still, like thunder, my body began screaming again. The pulse under my warm feverish skin, the aching in my lungs, the sore throat, the heaviness in my legs returned loudly and unexpectedly. Leaving this body was nice, but now I was back and I couldn't escape. I was stuck. Body, what are you trying to tell me? As I lay heavy, I wondered what my life would look like if I were to be sick like this forever. It wouldn't be a life, it would be a prison. I let myself go down rabbit holes of doom imagining Yiannis leaving me, my life falling even more to pieces. Anxiety began to rush into my chest, my arms and my legs. My heart beat faster and faster. I felt afraid, insecure, helpless. I didn't stay here long because fear turned into anger. A wave of rage and frustration swallowed me whole. This is not fair. I felt a tear roll down my cheek, soon my whole face was wet. I didn't bother wiping my tears. This is not fair!! Why me? That night I asked Yiannis not to carry me back into bed but to leave me to lie down on the carpet on the floor. I wanted to feel the ground beneath me. I wanted to feel my body raw and hear whether there was anything beneath the screaming. He very lovingly and supportingly stayed in the living room with me and took his place on the couch. Eventually we both fell asleep.*

Journal Entry: 25 June 2020

I just remembered something that I had forgotten. Two years ago, when my body aches began, I was listening to an episode of Dhru Purohit's podcast about the brain. It was about how the cells in our body are always listening to the thoughts in our mind. At that moment I was literally saying 'I am so tired' to myself. I remember realising that I said that to myself a lot. By the end of the episode I was convinced. I would change my thoughts and tomorrow I would never be tired again. I soon forgot about the promise I made to myself at Bayswater Station, and continued feeling exhausted. What occurs to me now is that maybe my mind was not speaking to my body – maybe it was the other way around. Maybe my mind was trying to listen to my body but it didn't know how. And maybe my body was whispering something that my mind couldn't understand. I think I'm onto something but I'm too tired to think.

For about five years, Jen would experience bouts of fatigue, pain, fear and panic attacks, and oscillate between high-functioning anxiety and generalised anxiety disorder. She would swing between perfectionism and overwhelm in an effort to cope. The toll of her dysregulation eventually produced aches and pains, waves of tiredness and back tension that became too loud to ignore. And because she didn't yet speak the language of the body, she kept pushing and surviving hoping it would all work itself out and one day go away. Spoiler alert: It didn't go away. It doubled down. Jen went from exercising six times a week, studying a degree in physiotherapy at a top university in London, seeing clients and patients before and after study hours, and working in the hospital ... to *nothing*. It was like the floor had been swept from beneath her feet. She became bedridden with what was diagnosed as chronic fatigue syndrome.

For many months it was an uphill battle trying to stick to physiotherapy rehab for chronic fatigue and talk therapy for anxiety. Sadly, nothing was *actually* working. Talk therapy triggered her physical symptoms and rehab triggered her anxiety. After resting in bed for nine months, she relapsed even harder. She was totally stuck and felt betrayed by her body. She felt lost. Jen loves to spend weeks deeply exploring subjects that interest her, so she dove into the realm of her body and finally stumbled upon a role of 'the nervous system' that she hadn't learned about in her biomedical science and physiotherapy studies. She discovered Porges' polyvagal theory and suddenly everything began to make sense. Her body wasn't suddenly weak and broken, it was trying to communicate with her. She learned that her nervous system was stuck and dysregulated, and that by regulating it she could heal herself.

Jen's dad, with whom she now has a repaired loving relationship, left when she was five leaving her feeling unwanted, inadequate, unloved and hurt. Really hurt. (*Hi Dad, if you're reading this book, I'm really proud of all the work you've done to heal yourself, it helped me heal too. Sometimes I still feel unwanted, inadequate, unloved and hurt, but then I remember you felt all those things too and didn't know how to love me and care for me like you would have wanted to.*) Jen's mother raised her alone and did the very best she could, loving her deeply. But Jen's mother was running on anxious patterns fuelling a mind and body with a lot of unhealed trauma and a dysregulated nervous system which made Jen feel perpetually anxious and dysregulated. (*Hi Mom, if you're reading this book I know you've had a lot of tough moments in this lifetime and I'm proud of you for being so strong and resilient. I'm sorry for the hurt you felt when I set a boundary between us. We have come such a long way to heal and repair our relationship. I am very grateful to you and I wouldn't be where I am today without you. I love you.*) These two dynamics created a nervous system that adapted with patterns of self-criticism, self-doubt, perfectionism, people pleasing, fear, freeze and overwhelm. Jen experienced years of anxiety, stuttering, fatigue, aches and pains, and all this time it was her nervous system - stuck in survival mode asking for appropriate soothing and healing. Jen was stuck in freeze and fight and flight for a long time. Chronic fatigue was just the last straw to break the camel's back.

After becoming aware of her nervous system states and of her mind-body-human messages, interrupting the nervous system patterns that were keeping her stuck and redesigning her neural circuitry for healing and regulation, Jen fully recovered. She has co-written this book for you so that you can heal too. And the real takeaway, aside from the unwavering power of self-healing that exists within us, is that our body is always communicating messages. Some feel good and some don't feel good. These messages, signals and wisdom are being communicated by your nervous system through your physiology. Behaviours like procrastination, perfectionism, people pleasing and over-achieving, and symptoms like back tension, neck pain, stomach ache, CFS and IBS are screams on top of the whispers of unresolved emotions and sensations like stress, fear, anger, numbness, loneliness and sadness that you didn't know how to hear. When we don't know the meaning of the whispers, the body begins to shout. Jen likes to say that chronic illness is the body's way of shouting for help. In this chapter we are teaching you a technique that allows you to translate what nervous system states you are in so you know what to do with what you experience. In the awareness practice of Part 1, you practised listening techniques. Here, you will learn to translate those messages from your nervous system through the polyvagal system.

Rethinking Diagnoses

> Through the lens of the polyvagal theory, almost every diagnosis in the DSM [Diagnostic and Statistical Manual of Mental Disorders] is a dysregulated nervous system.
>
> Deb Dana, LCSW, author of the *Polyvagal Theory in Therapy* and *Polyvagal Exercises for Safety and Connection*

For those of us that have been suffering from the impacts of nervous system dysregulation, it's confusing and overwhelming because we don't know what is wrong with us. Like Jen, we may feel stuck and frustrated. In the beginning, without truly understanding how our nervous system is involved, we just know that we experience persistent symptoms that are often given labels like depression and IBS. Although these diagnostic labels sometimes come with a sense of relief (because having a name for something, even if it's wrong, gives us a sense of understanding) they may not always help. Often diagnostic labels pathologise us - meaning they reinforce the perception that we are sick and broken, which adds to our anxiety and depression, fuelling our fear and the dysregulation of our nervous system.

When you come to realise the central role of the nervous system in your body and then translate it through the lens of polyvagal theory, everything that is happening to you begins to make perfect sense as something that is happening *for* you. To keep you alive and help you survive. This may seem revolutionary and there will be sceptics, but when we look at the body and how these exercises have helped so many of our clients heal and achieve more calm and regulation in their nervous systems, we can see that working with these practices is truly revolutionary. When we apply the perspective of nervous system dysregulation to depression and IBS, we see it as someone whose nervous system has become stuck in a dorsal vagal state causing them to exhibit psychological symptoms like dissociation, sadness and hopelessness ('depression') plus the biological symptoms of decreased gut function and low energy ('IBS'). The treatment would be to use mind and body nervous system regulation techniques to bring them up out of the dorsal and back into harmony with their sympathetic and ventral vagal states.

In the old paradigm, when seeking out help, you got a diagnosis of 'major depressive disorder' or 'burnout' and 'irritable bowel syndrome' and were sent to a talk therapist and gastroenterologist for completely independent and non-integrated treatment. In the new paradigm, we see the two diagnoses not as distinct disorders to be treated separately, but as two labels describing symptoms of the single actual diagnosis of a dysregulated nervous system. By placing our therapeutic interventions on the nervous system with A I R, both the depression and the IBS - along with other ailments - can be repaired together for the long term. This has been the experience of thousands of our clients who have fully recovered.

Alyssa *reached out to Jen on social media two years ago. She had been struggling with a number of symptoms for several years. It began with episodes of dizziness, rapid heartbeat and an overwhelming feeling of weakness every time she stood up. After numerous doctor visits and misdiagnoses, Alyssa was eventually diagnosed with postural orthostatic tachycardia syndrome (POTS). This diagnosis left her feeling scared, and uncertain about her future. Alyssa became Jen's client. Over the course of five sessions, they focused on cultivating Alyssa's awareness of what polyvagal state she was getting stuck in. Alyssa noticed a lot of anger inside her and that her body felt it was ready to lash out at everything, and realised she was nearly always in an agitated sympathetic state. She learned to interrupt the sympathetic state she was stuck in through vagus nerve and somatosensory techniques. As her triggers would initiate a rise in sympathetic activation, she would use her exercises moments before an expected heart-rate increase or dizziness spell would happen and she stopped the symptoms before they could begin. The process of A I R allowed her to change the messages coming from her body into her brain and then rewire those signals travelling back down from the brain to the body. The new messages helped lower the anger–fear–symptom feedback loop and switched off her overactive sympathetic nervous system while switching on her more regulated ventral vagal state.*

With each session, Alyssa began to uncouple her attachment to her diagnosis and realise that the state in her physiology was changeable. She also began learning that her body had been asking her to slow down for a long time, giving her hints through random episodes of irregular heartbeats and dizziness. Alyssa learned to connect with herself in a way she never had before. Her anger which had been repressed for so long was being felt and discharged. And she found herself feeling relief for the first time in years. Jen remembers a particular moment that Alyssa shared with her via DM. After using tapping, vocalisations and vagal toning she could stand outside and enjoy the warmth of the sun on her skin without it causing her heart rate to abnormally increase. 'Being able to stand in the sun reminds me I am not broken. It helps me feel normal again and feeling normal gives me my power back. I will never forget that I have everything I need inside me to heal myself and I can't believe it was there all along. This work has changed me forever. Thank you.'

As Alyssa continued to practise A I R in her daily life, her recovery progressed rapidly. Her symptoms gradually subsided, and by the end of her sessions with Jen, she was virtually symptom-free. She was able to stand up without dizziness, engage in physical activities and enjoy a newfound sense of self-empowered well-being. Alyssa's journey from a place of desperation and fear to a full recovery was a testament to the power of the effectiveness of nervous system techniques. Her story, shared by Jen on social media, offered hope to others who were just like Alyssa.

Loss of Will or Dorsal Shutdown

One summer Karden had a client named **Dave** *come to his office who had spent the last few years fighting and miraculously overcoming a brain tumour. Before the brain cancer, by all reports, Dave had been an outgoing, personable and highly social person who could light up a room. In the year since, the cancer had been eradicated and most of the side effects of the interventions had worn off, but Dave was not himself. His indomitable spirit had been muted, he was struggling to make progress with the various rehabilitation treatments he was receiving and completing everyday tasks had become, in his own words, daunting. He had been to the best specialists in the world to structurally and functionally evaluate his brain and nervous* *system, and though there was some brain damage from the tumour and interventions that accounted for some of his physical difficulties, they certainly did not account for his psychological and emotional deficits. In addition, he had been working with a smart and experienced psychoanalyst for over a year without much progress.*

Immediately upon seeing him, through the lens of the polyvagal theory, Karden could see that Dave was predominantly in a dorsal vagal state. This was communicated by his hunched posture, lowered head, quietude, dim eyes and overall impression of collapse and shutdown. He was in what Peter Levine would call a 'functional freeze' – capable of living and surviving each day but with no real life, joy and vitality. In order for Dave to get out of dorsal shutdown, Karden needed to help Dave activate his own sympathetic survival energy. To do that, Karden first led him through an embodied visualisation of a memory from when Dave played soccer. In the memory he could sense the grass under his feet, the muscles of his legs flexing, his heart pumping and the camaraderie with his teammates. As he summoned this memory, subtle changes in his posture and facial expression indicated that he was beginning to feel some strength and vigour in his body, which signalled to Karden that his nervous system was gently moving from dorsal shutdown to sympathetic activity. Then Dave and Karden transitioned to a boundary exercise. Karden stood 10 feet away and told Dave that he was going to take slow steps towards him and that whenever he felt like he wanted Karden to stop, to say 'STOP' in a clear voice. Maintaining boundaries is an important part of keeping ourselves safe and the sympathetic nervous system is the gatekeeper. By asking Dave to become aware of his boundaries and enforce them by saying 'STOP', his sympathetic nervous system became even more invigorated. Like a match igniting a candle, Dave's sympathetic nervous system came online and spontaneously shifted him up the hierarchy, sending him out of shutdown and back into aliveness.

Though illustrating how someone can move from one polyvagal state to another is critical, it is not the lesson of this story. The lesson is found in comparing the approach of Dave's psychoanalyst and Karden. In his notes, the analyst said: 'I am accustomed to seeing patients who don't allow themselves to hope or to believe in themselves and their abilities. For most, these fears are unrealistic and psychological in nature. But Dave's encounters with death have compromised his ability to have hope and faith in his own resources, in a way that is really hard to understand for those of us that have not been so terrifyingly exposed to our mortality and limits of our capacities. This is the challenge of Dave's situation.' These notes are fascinating because they tragically and accurately describe Dave's circumstances and the therapist, seeing through his lens of psychoanalysis, is arriving at a similar conclusion as Karden. Dave is shut down. But the most important difference is that the psychoanalyst's observations, though accurate, and his therapy sessions with Dave, though compassionate and insightful, hadn't significantly helped Dave to return to his old self. And that's because the secret language of the body doesn't speak words of psychoanalysis, it speaks in BASE.

It's not that Dave wouldn't allow himself to hope or believe in his own capacities or that he was reminded (at least unconsciously) of his fragility and limits through his near-death experience with brain cancer. It was that his three-tiered polyvagal-based neuroceptive survival system, faced with the horrifying and overwhelming threat of brain cancer, a thing it couldn't fight or flee, activated the only survival pathway it had left, the dorsal vagal response, and got stuck there. The session he had with Karden worked because it was a non-verbal conversation at the level of neuroception and it allowed Dave's nervous system to find its way back to sympathetic and ventral states. Once that happened, Dave regained access to his own 'capacities' as well as his zest for life. Through a combination of subsequent sessions and practising A I R, Dave regulated his nervous system and by doing so his mind and personality were restored as well.

> **It's not in your head, but it is in your brain.**

In the polyvagal and trauma-informed nervous system framework, diagnoses like 'generalised anxiety disorder' or 'adjustment disorder with depressed mood' are simply labels describing a collection of symptoms. When viewed through the lens of the nervous system and worked on through the language of the body, not only do these disorders make sense, they can be healed and reversed.

You are not your diagnosis, there isn't something wrong with you, you don't have a defect, you are not broken and it's not that you simply aren't trying hard enough to be happy or to be calm. The truth is that for no fault of your own, your nervous system is dysregulated and trying to 'fix it' from the mind can't work because the executive functions of your mind *can't function* on top of dysfunction of your hypervigilant surviving nervous system. Just like Dave, it's not that you aren't allowing yourself to believe in your capacities or that you need to just change the way you think and feel, it's that your polyvagal system is stuck in a place that makes it challenging for you to shift things. But we see this as an opportunity. One where you finally feel that you have a choice over your health; the choice to regulate your nervous system, strengthen its capacity for resilience and finally heal yourself.

A Review of Nervous System States

Hopefully you are now beginning to see that your body provides subtle warnings before it raises a loud alarm, and by *translating* the early cues and messages of your survival responses through BASE-B, you can often prevent it from sounding that alarm. Through the trauma-informed lens of the polyvagal theory, your nervous system adopts a hierarchy of responses to help keep you safe. These responses correspond to nervous system states, physiological adaptations, and physical, behavioural and emotional presentations. See below how these can be summarised.

1. The **ventral vagal state** is associated with feelings of safety, social engagement and relaxation. *You know when you are in this state because you experience a felt sense of calm, emotional connection and overall well-being.* Physiologically, it promotes a normal heart rate, relaxed muscles and deep, rhythmic breathing. Chronic stress, trauma or disrupted social connections can potentially lead to a shift away from this state. Chronic illnesses linked to a lack of ventral vagal balance may include anxiety, depression and other mood disorders.
2. The **sympathetic state** is activated in response to stress, perceived threats or challenges. It is characterised by heightened alertness, increased heart rate, muscle tension and a readiness for action. *You know you are in this state when you experience high energy rushing in your arms and legs, anxiety, fear, irritability and restlessness.* Chronic activation of the sympathetic state can lead to conditions like hypertension, heart disease and chronic stress-related disorders.

3. The **dorsal vagal state** is associated with immobilisation and a sense of disconnection. It is characterised by emotional numbness, physical immobility, dissociation and reduced heart rate. *You know you are in this state when you experience low mood, difficulty concentrating, mind and body numbness and social withdrawal.* Chronic or extreme stress can lead to a chronic dorsal vagal state, contributing to conditions like depression, fibromyalgia and other stress-related illnesses.
4. The **freeze state** is the simultaneous activation of the sympathetic and dorsal states associated with a sense of physical immobilisation in combination with a highly charged amount of energy – much like how it would be to press the gas and brake pedals of a car at the same time. It is characterised by emotional numbness, physical rigidity, immobilisation and a lowered heart rate. *You know when you are in this state when you experience feeling stuck in all of or a certain part of the body, always feeling cold or numb, physical stiffness or heaviness of limbs, restricted breathing or holding of the breath, a sense of dread or foreboding and overwhelm.* Chronic stress and unresolved trauma can lead to a chronic activation of the freeze response causing a wide range of nervous system dysregulation symptoms like chronic fatigue syndrome, fibromyalgia, anxiety and depression.

Chronic activation of any of these states is a sign of a dysregulated nervous system, which can be repaired and healed through learning the secret language of the body. Let's start *translating*!

Let's take a moment to summarise the knowledge on *translating* the messages for awareness from your body:

- Our nervous system is governed by keeping us safe.
- Neuroception is the survival-driven, non-verbal, automatic and instantaneous conversation and actions being carried out by your brain and body beneath your typical cognitive perception.
- The polyvagal theory says that, when exposed to a threat, your nervous system will deploy a three-tiered hierarchy of responses to keep you safe.
- Getting to know your nervous system states is what the exercise of *translating* will teach you.
- Changing the dysregulated conversation between your nervous system and your body is possible and can bring harmony to your life.

Practices 4

Awareness: Translating

Learn to translate the messages coming from your body into the survival language of your nervous system

In these practices you will learn how to recognise the specific nervous system survival state you're in through BASE-B.

Why Do It?

The translate techniques will help you learn and understand that you do not just have depression or anxiety, you have a nervous system that is stuck amidst the spectrum of polyvagal-based survival states.

When to Do It

Use translating techniques when you experience a trigger and to go beyond basic listening to determine what survival pattern is underlying your dysregulation. For example, if you are healing from pain and have used listening to notice the tension and anger alongside the pain, you would use translate to learn that you may be stuck in sympathetic fight mode. Doing this will help build awareness and inform what techniques to use to regulate yourself.

Tips before you begin

- Reminder: YOU ARE NOT BROKEN. Even if your nervous system employs default patterns that feel too strong for you to influence, remember that you are only in step 'A' of A I R in the body. Healing happens in steps, and this is only step one.
- The survival states that you will be learning to identify and the messages coming from your body are often easy to determine. But sometimes you may wonder what state you're in if the body is sending you many different messages. This is normal. The goal of this work is to become familiar with the strategy your nervous system employs. It's not a perfect process with a perfect outcome. It's a work of exploration, enquiry, curiosity – and self-care.

Practice 1: Learning the Body's Responses

1. Find a quiet place to sit.
2. Take a deep breath.
3. Think of something that *mildly* triggers you. For example, public speaking, your boss giving you negative feedback or your inbox full of emails.
4. Notice the first thing that happens in your body. For example, a sudden feeling of achiness in your throat, an electric-like feeling in your heart making it pump faster and louder, blood rushing through your arms and legs, a feeling of numbness and struggling to connect with any bodily sensation at all.
5. Resist the urge to fix or change anything, and stay here with what arises a little bit longer.

6. Now that you're aware of the messages coming into your mind, begin listening to them individually through BASE-B and label them:
    a. **Breath:** What am I feeling in my breath?
    (I am feeling my breath shorten and quicken, my chest feels tight.)
    b. **Action:** What does my body want to do and how does it want to move?
    (My body wants to move, I think it wants to punch or kick something.)
    c. **Sensation:** What am I feeling in my body and where?
    (I am feeling energy and blood rushing into my arms, legs and face. I can feel my heart rate increasing.)
    d. **Emotion:** How am I feeling and where?
    (I am feeling anger in my arms and hurt in my chest and belly.)
    e. **Body:** What polyvagal state am I in?
    ('I am in sympathetic fight mode.')

Repeat this practice a few times on the same trigger until you have a description of the state(s) that you are in. Then compare your description to the guide below (pages 151–7) to translate your experience into the survival state you are in. If you sense you are in a blended state, that is OK too!

Then, use this exercise on other triggers to learn even more about your body and explore what other survival patterns feel like.

Practice 2: Learning Your Body's Default Response

1. Find a quiet place to sit in.
2. Take a deep breath.
3. Think of what you are reading this book for. Choose one thing to begin with. For example, anxiety, depression, unhealed trauma, chronic pain or burnout.
4. As you think about that thing, notice the first thing that happens in your body. For example, a sudden feeling of heat on your skin, a skipped heartbeat, an empty feeling in your gut, a feeling of being outside your body or a strong sense of fear rushing through your entire self.
5. Resist the urge to fix or change anything, and stay here with what arises a little bit longer.
6. Now that you're aware of the messages coming to mind, begin listening to them individually through BASE-B and label what you notice:
    a. **Breath:** What am I feeling in my breath?
    (I am feeling my breath shorten and quicken, my chest feels tight.)
    b. **Action:** What does my body want to do and how does it want to move?
    (My body wants to move, I think it wants to punch or kick something.)
    c. **Sensation:** What am I feeling in my body and where?
    (I am feeling energy and blood rushing into my arms, legs, and face. I can feel my heart rate increasing.)

d. **Emotion:** How am I feeling and where? (I am feeling anger in my arms and hurt in my chest and belly.)
e. **Body:** What polyvagal state am I in? ('I am in sympathetic fight mode.')

Repeat this practice a few times on the reasons why you are reading this book until you have a description of the state(s) that you are in. Then compare your description to the guide below to translate your experience into the survival state you are in.

You have now identified your body's default survival response. In knowing this, you'll be able to choose the appropriate A I R techniques to help you shift out of your default survival state and into a state of regulation.

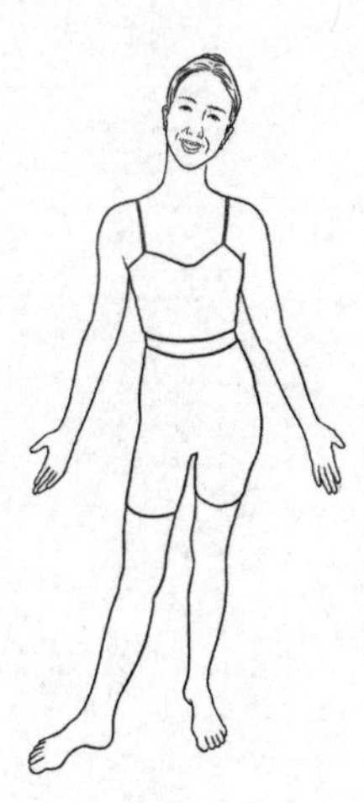

Ventral vagal mode

- Regulated
- Adaptable
- Flexible
- Resilient
- Settled
- Calm
- Grounded
- Curious
- Trusting
- Receptive
- Open
- Compassionate
- Caring
- Optimal organ function
- Increased heart rate variability (HRV)
- Increased recovery
- Better immunity
- Vitality

- Social
- Interactive
- Collaborative

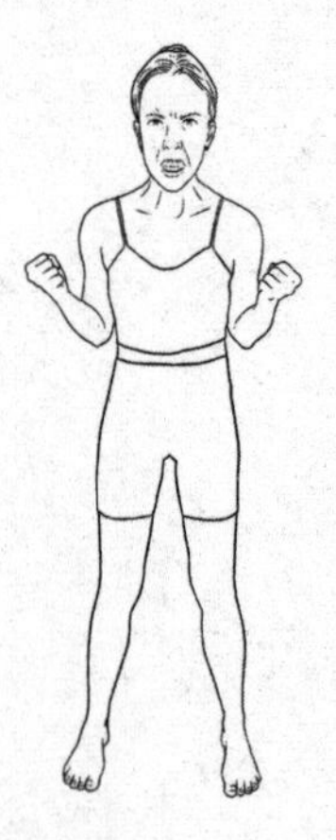

Sympathetic fight mode

- Dysregulated
- Hyper-aroused
- Sensitised nervous system
- Anger
- Irritation
- Frustration
- Annoyance
- Hypervigilance
- Anxiety
- Increased heart rate
- Muscle tension
- Gastrointestinal problems
- Headaches
- Migraines
- Compromised immune system

- High blood pressure
- Weight gain or loss
- Decreased libido
- Insomnia
- Increased sensitivity to stimuli
- Aggressive behaviour
- Difficulty concentrating
- Memory problems
- Social withdrawal
- Reduced productivity
- Cardiovascular diseases
- Hyperglycemia
- Kidney disease
- Cancer
- Chronic pain

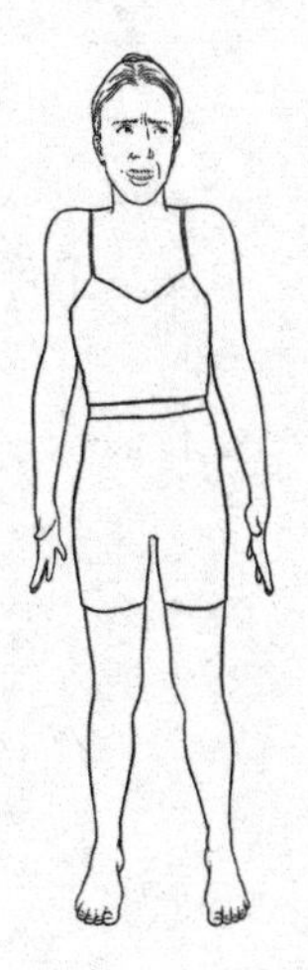

Sympathetic flight mode

- Dysregulated
- Hyper-aroused

- Sensitised nervous system
- Panic
- Fear
- Worry
- Anxiety
- Hypervigilance
- Increased heart rate
- Muscle tension
- Gastrointestinal problems
- Headaches
- Migraines
- Compromised immune system
- High blood pressure
- Weight gain or loss
- Decreased libido
- Insomnia
- Increased sensitivity to stimuli
- Aggressive behaviour
- Difficulty concentrating
- Memory problems
- Social withdrawal

Dorsal shutdown

- Dissociation
- Derealisation
- Numbness
- Depression
- Raised pain threshold
- Helplessness
- Shame
- Shutdown
- Emotional numbness
- Social isolation
- Apathy and indifference
- Hypoarousal/feeling mentally and physically under-aroused/disengagement
- Inability to focus or concentrate
- Loss of interest in hobbies and activities
- Difficulty with decision-making
- Physical discomfort/chronic physical tension or discomfort/pain

- Memory problems
- Procrastination
- Sleep disturbances
- Changes in appetite
- Feelings of hopelessness

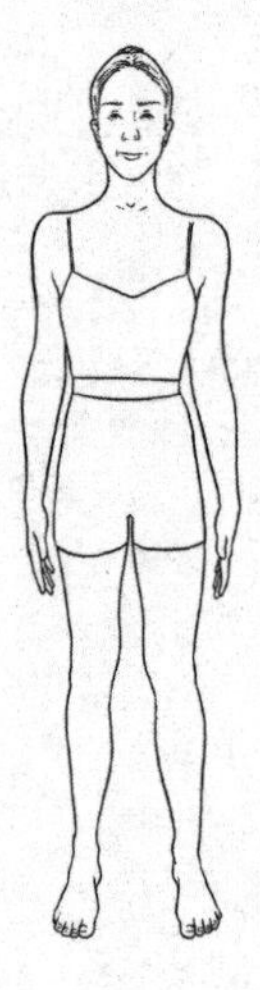

Dorsal freeze mode

- Emotional numbness
- Depersonalisation
- Difficulty in decision-making
- Loss of motivation
- Social withdrawal
- Apathy
- Physical discomfort/chronic physical tension, often leading to discomfort and bodily pain
- Difficulty with speech
- Memory problems
- Procrastination
- Feeling overwhelmed

- Avoidance behaviours
- Decreased libido
- Inability to set boundaries
- Intrusive thoughts
- Hypoarousal
- Lethargy
- Changes in appetite
- Sleep disturbance
- Low energy
- Feelings of hopelessness

Make Translating a Part of Your Life

The goal is to consistently see through symptoms and identify the underlying survival response that's driving your dysregulation so you can take effective action. To do this, we invite you to:

1. Remember that YOU ARE NOT BROKEN.
2. Always presume that underlying your unhelpful thoughts, body sensations, behaviours and symptoms is a survival state.
3. Instead of trying to fix your unhelpful thoughts, body sensations, behaviours and symptoms, your goal is to use your Observing Self to identify the underlying state and use A I R to shift it into a more safe and regulated state.

Notes:

....

....

....

....

....

....

Chapter 5: Interruption

Modifying

the responses in your nervous system to achieve regulation

Avoiding your triggers isn't healing. Healing happens when you're triggered and you're able to move through the pain, the pattern, and the story and walk your way to a different ending.
Vienna Pharaon

Now that you have learned about *translating* the secret language of your body, you are ready to actively interrupt the dysregulated conversation with the technique in A I R called *modifying*. We call the modifying techniques NSMs (nervous system modifiers) and they are designed to help you move your nervous system out of the patterns it has become stuck in and return it to adaptability and flexibility. You will learn a variety of NSMs designed to help in many situations. Some are designed for soothing the sympathetic fight-and-flight response, others are for gently mobilising you out of the freeze response of the dorsal vagal state while still others are geared toward the overall 'toning' and health of your vagus nerve. As with the other practices, you will begin to recognise which ones work best for you and in which specific situations.

To help you better use these techniques we are going to spend some time in this chapter discussing why your nervous system gets stuck in these states, describe how a regulated nervous system functions and explore the fascinating anatomy and physiology of the vagus nerve in order to understand why NSMs are so effective at helping you regulate. As we emphasised in the last chapter, the dysregulated state of your nervous system is not your fault. It's not because of a defect or a lack of willpower, it's because certain vulnerabilities exist in how our brains and bodies try to use our intrinsic survival mechanisms to navigate stressful and traumatic events. And although you are not responsible for your current state of dysregulation, once you have the knowledge and skills you are gaining from this book, you can choose to be responsible for leading yourself back into regulation.

It is worth noting that while we are self-healers, we don't need to heal alone. Taking responsibility for leading your nervous system back into regulation does not mean 'it's all on you' or you need to 'do it alone'. In fact, we encourage you to seek as much support as possible from family, friends and professionals as you tailor these practices to you. Just remember that no matter how big your team of support is, when it comes to healing, it is most helpful for you and the observing mode of yourself to be the leader of the team and your healing. Here are a few of the notes and messages from our clients on how their work with NSMs has helped change their lives:

> Oh my goodness. I just tried the vagal toner and I cried for 20 minutes. I feel so much ease and rest in my body. I didn't know such a simple movement could be so powerful. Please keep sharing these, they are changing my life.
>
> Anna

This tapping exercise is incredible. So beautifully articulated and explained and for the first time in my life I feel like I have something tangible to help me heal. I've been doing your practices for a few weeks now and my anxiety has improved so much already. I am shocked by how quickly I feel better.

Jo

I thought I would reach out personally. I am a physiotherapist and clinical lead in the London NHS and a patient referred me to your social media to ask me what I thought about the exercises you teach. I have had my own curiosities about this work for some time now but I never dared steer away from what I was taught (because that is how I was trained). I looked at all of the content you shared and read the testimonials and then started using some of your vagus nerve exercises and nervous system-based techniques (vocalisations and tapping). I must learn where you learned all of this because - aside from it being logical and science-proven - it is extremely effective. I unfortunately experienced long Covid and got stuck trying to pace myself (as a good physio), only to get worse. I took a few months off and started meditating and being more conscious about my thoughts and unknowingly 'regulating my nervous system' as you say ... but wow these vagus nerve modifiers are powerful. Thank you for being brave and paving the path forth in a world that desperately needs answers and healing.

Nate

Seriously, I have tears in my eyes and a lump in my throat. I wish I had social media as a young man. Luckily I'm on the other side, but I have a 24-year-old son who was struggling immensely with chronic pain. I sent him your exercises and since he has been doing them his pain has diminished. What a lucky time to be alive to be able to have access to this invaluable information!! I wish my doctors had shown me these back when I was struggling with depression. Thank you for all that you do and I am grateful to have found you ... better late than never.

Andrew

I've been in your programme for a month now and your vagal techniques are simply like nothing I've seen before. I use them to help me when I experience a flare in chronic fatigue and together with the other tools I am recovering. I can't believe I'm saying this, but I am actually recovering. Who knew that I had all the answers inside me after years searching for answers outside of me.

Helen

Thank you for pushing the message of healing the nervous system and for reinforcing it with your caring, accessible and gentle style. All of the NSM exercises have allowed me to be less afraid. I can instantly feel a shift in my body and I know that it's working. I am so proud of my body for being fiercely strong. I am learning that I am resilient, thank you for giving me this gift. I am beginning to feel safe in my body for the first time since I can remember.

Claudia

I am in awe of the vagal exercises. So easy to work through and as a psychoanalyst and someone who has experienced anxiety and depression, I am going to start using these on myself when what I do doesn't work. Believe it or not, I didn't learn any of this in school! Seeing the recovery success of your clients is astounding, I can't wait to learn more about all of this somatic work so I can start using it with my clients too.

Linda

These comments reflect the collective sentiment of our thousands of clients who have begun using NSMs to modify the responses of their nervous systems to heal themselves.

Pattern Recognition and Habit Repetition Machines

Though we like to think of ourselves as conscious, sentient and thoughtful human beings living our lives with full awareness, discernment and intention, this perception is mostly unrealistic. We are creatures of habit. Most of our thought processes are automatic, not intentional. We make many of our decisions throughout our day using mental shortcuts. We wake up, hit the snooze button, hit the snooze button again, roll out of bed, brush teeth, get coffee, commute to work along the same path, solve the usual problems at work, have the same small talk with our colleagues, answer the same questions that our mother messages us, commute home along the same path, eat one of six or so favourite meals, watch a show on Netflix, brush teeth again, doom scroll our phones for two hours, fall asleep. Repeat. We are in fact, for the majority of our waking life, pattern recognition and habit repetition machines. And this isn't because we live in the Matrix, it's because our brains are designed to be efficient and save cognitive space for higher-level problem-solving.

In the words of Charles Duhigg, the author of *The Power of Habit*: 'the brain is constantly looking for ways to save effort. Left to its own devices, the brain will try to make almost any routine into a habit ... this effort-saving is a huge advantage.'[1] The advantages come in many forms. The first is energy efficiency. Your brain is arguably the hungriest organ in the body, consuming roughly 20 per cent of your energy each day.[2] Habits allow the brain to devote fewer resources and consume less energy when trying to execute a task. Another big advantage of habit is speed. The ability to do things fast and efficiently, especially survival instincts/habits, greatly increases the chance that you'll survive a dangerous encounter. When it comes to evolution and survival of the fittest, being quick was the difference between eating dinner or being dinner. Another benefit is that when we can automate a task into a habit, like washing the dishes, we can think about other things while we do the task. We could go on but we trust that you can understand the crucial role habits play in our lives and why the brain is so good at creating them.

All habit loops consist of three essential components: a cue, a routine and a reward. Making coffee as soon as we wake up is a good example. Another example is your morning routine. If Karden sets his morning alarm at 6 a.m. to exercise (cue) he will then change into his workout clothes, go to the gym, plug in his headphones and exercise (routine) and experience the endorphin-filled satisfaction of feeling strong and healthy (reward). Unfortunately, when it comes to nervous system dysregulation, our powerful autonomic responses to chronic stress and trauma combined with the brain's propensity for habit formation become formidable obstacles in your efforts to heal. The intensity and repetition of stressful and traumatic events (that form cues) and the corresponding coping mechanisms (that form routines) deployed by the nervous system to survive (reward) become strongly ingrained in the parts of the brain that establish habits. When it comes to our lives, cues are felt as triggers, routines are the familiar symptoms of anxiety, hypervigilance, worry, depression, numbness and shutdown, and the reward, well, it doesn't feel like a reward, but as far as your nervous system is concerned, not dying is a win.

Despite the strength and intensity of these survival habits, they are not carved in stone. These habits are built within your brain tissue, which is quite malleable thanks to neuroplasticity. This means that they can most certainly be changed but one must employ the right combination of strategies to do so. We've already laid the groundwork for you to be able to do this by teaching you how to use A I R through listening, switching, distancing and translating and we are going to take it to another level with modifying.

Although nervous system regulation is more than just working with habits formed in your neural pathways, using the science of habit transformation plays a critical role in healing. The strategy behind modifying the dysregulated habits of your nervous system involves two components: the first is interrupting habit loops and the second is inserting new routines to replace the old ones. Most of us have had experiences interrupting habits and forming new ones, even if we were doing it subconsciously. Travelling to a foreign country tends to be a marvellous habit interrupter. Since so much is unfamiliar to us, the brain can't recognise many of its typical cues and therefore has difficulty executing typical habits. This is why aspects of your day will seem more vivid while travelling, more delicious, more beautiful and more meaningful. When you are abroad your brain becomes deeply engaged with your daily experiences, rather than bypassing them with habits. This is also part of the reason travel can be so exhausting and causes some people anxiety. Your mind and body are out of their comfort zone.

Habit replacement also happens frequently. For example, Karden had been eating a fried egg on toast nearly every morning for an entire year. One day his wife suggested having hard-boiled eggs for a change. Then, for the next several months, Karden completely forgot about his previous routine of fried eggs and started eating hard-boiled eggs every morning. Habits are fascinating because though they can be very tenacious, they can also be swapped and changed quite easily.

With modifying, you will begin to interrupt and replace the dysregulating routines of your nervous system with the regulating routines of the NSMs. It's important to note that the goal isn't to not get triggered (though over time by applying this work your triggers will become less and less intense) but instead to recognise the triggers as early as possible by listening to the language of your body, observing it, translating it and replacing the old unhelpful survival routine with a new NSM or other regulating routine and doing so with consistency until the new habit forms. The old habit never really goes away, but like a trail through the forest, the less and less it's used, the more it becomes overgrown and fades away.

> **To heal, your nervous system needs you to practise replacing old patterns with new ones, over and over again.**

Window of Tolerance

When it comes to regulating our nervous systems, the habits we're trying to change are not obvious behaviours like eating too much ice cream, avoiding exercise or spending too much time on screens. Our goal is to change the more elemental, neuroception-level habits of our stress response and survival system. Believe it or not - although it is not the goal of this book - if you can reform the survival habits and begin *settling* (next chapter) your nervous system, many typical undesirable habits (like binge eating) often resolve themselves. This is because many of our less helpful behaviour habits are actually coping mechanisms designed to resolve or avoid our underlying feelings of dysregulation. If you heal the dysregulation, the coping pattern is no longer triggered/needed. This is actually also true for symptoms. When Jen had chronic fatigue syndrome she also broke out in severe acne, for the first time in her life. While it was frustrating and triggering, because she was bedridden, the acne was the least of her worries and problems. As Jen began to heal herself and modify the responses in her nervous system, as well as going from not being able to walk to the bathroom to walking a daily 10k, her acne completely disappeared. She wasn't trying to make it go away. She didn't use any products, supplements or skin treatments - she regulated her nervous system. In case you're curious here are other symptoms that *completely* disappeared: painful and heavy periods to the point of nausea and vomiting, hair loss, widespread blotchy red skin, unexplained nausea, visual snow, eye floaters, twitching in fingers and leg muscles, chest tightness, constant tension headaches, ectopic heart beats, unexplained fevers, chronic sore throat and swollen lymph glands. Symptoms are messages in the language of the body communicating that there are unresolved feelings perpetuating unhelpful survival responses. Speaking the language of the body gives you access to *modifying* these responses and, by doing so, healing yourself.

As we reviewed in the last chapter, according to polyvagal theory, when the nervous system detects a threat it will deploy a three-tiered hierarchy of adaptations. Our nervous system becomes dysregulated because it loses the ability to fluidly transition among these states and the emotions reflected in these states, and instead gets stuck in extreme sympathetic or dorsal vagal reactions or see-saws between the two. A model for describing this transition is called the 'window of tolerance'.

This concept was developed by Dan Siegel, a clinical professor of psychiatry, in his 1999 book *The Developing Mind*. According to Siegel, each person has a unique capacity for handling and making sense of their emotions. This capacity is often referred to as their window of tolerance or optimum arousal zone. The size of this window varies from person to person. Some individuals have a broad window of tolerance, allowing them to comfortably experience a wide range of emotional intensities, including both positive emotions like excitement and happiness, as well as challenging emotions such as guilt or anger. In essence, it's about how much emotional, physical and physiological intensity and diversity a person can navigate without feeling overwhelmed, shutting down or getting sick.

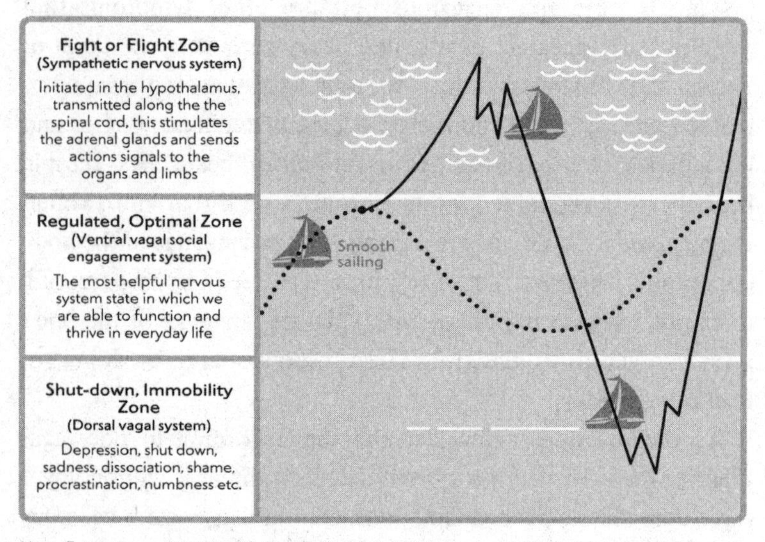

Note: *Freeze* is a combination of *fight or flight* and *shutdown*, where we have a build-up of energy but also experience numbness and imobility.

Everyone's window of tolerance – or zone of optimal arousal – has an upper and a lower boundary. Above the upper boundary is a zone of 'hyper-arousal' (sympathetic activation), and below the lower boundary is a zone of 'hypo-arousal' (dorsal activation). When our level of arousal moves beyond the boundaries of our window of tolerance we experience nervous system dysregulation and symptoms. Siegel's concept of the 'window of tolerance' can therefore be directly linked to the autonomic nervous system's functioning. Beyond the upper boundary of this window, there's an excessive activation of the sympathetic branch of the autonomic nervous system. This results in heightened energy consumption and a state of hyper-arousal. Below the lower boundary, excessive activity in the parasympathetic branch takes over, causing a decrease in physiological processes like heart rate and respiration. This, in turn, is accompanied by psychological experiences of feeling numb and shutting down, leading to hypo-arousal. Siegel also suggests that there are other potential combinations of sympathetic and parasympathetic activity. The middle zone, or optimal zone, is the ideal place to spend most of your time. Here, we are primarily living in a ventral vagal state which allows us to feel comfortable in ourselves and connected to people. As situations become more demanding, we are able to oscillate up towards the sympathetic and, as circumstances settle, we can oscillate down toward the parasympathetic.

To look at this in real life, you can imagine yourself out for dinner with friends. Ideally you are having a good time and chatting with your buddies (ventral vagal). Then your phone rings and you see that you have a message from your partner that says, 'Do you have a minute to talk?' Almost instantly, you become a bit more activated, your body becomes stiffer, your heart rate increases and you politely excuse yourself from the table (sympathetic). When you call your partner they tell you that your plane tickets for your holiday leaving the next day have been cancelled. The two of you problem-solve and agree to buy new tickets later that night. Your sense of initial fear followed by frustration begins to lessen, you take a deep breath and return to dinner with your friends and are able to re-engage and enjoy yourself (sympathetic back to ventral). In fact, you end up enjoying the meal so much that you feel the sluggishness of a food coma coming on as your body redirects its energy to digest the meal (dorsal vagal).

When you have a dysregulated nervous system, you can no longer easily transition between the various levels of arousal created by the stressful input and the dinner scenario goes differently. If your nervous system's habit is to overactivate the sympathetic response, when the text comes from your partner, the intensity of the activation is much higher, your body becomes very stiff, your heart pounds in your chest and you may feel nauseous. You immediately catastrophise and think they may be calling because someone has been hurt or to break up. You then abruptly excuse yourself from the table and fumble with your phone to call your partner. During the call, you are pacing on the sidewalk outside the restaurant, the conversation escalates your arousal, you ask dozens of questions and start blaming your partner for the cancellation, even though logically you know they have nothing to do with it. You feel extremely activated and don't even hear your partner trying to support you and help you out of this feeling. On the other hand, if your nervous system's habit is to go towards the dorsal vagal response, after you see your partner's message you may feel a spike of fear followed by your body becoming numb. You will feel as if your body is leaving the table and the restaurant even though you haven't yet moved. You shut down, become immobile, feel overwhelmed and trapped. You then don't call your partner because you are too frozen by fear of what that call might be about and instead completely disengage from the conversation and go inward, cut off from the outside world.

Reflecting on this example, if you have a frustrating last-minute change of plans, it's functional and useful to have a certain amount of sympathetic activation to address and solve it. But it's not functional or useful to be so activated that your nervous system sends you into panic, freeze or collapse. After you've successfully solved the problem, it's important that your system knows how to return to a baseline ventral state so that you can be comfortable in your body, connected to others and enjoy your life.

The entire purpose of *modifying* is to notice when these habitual survival loops are being initiated by your nervous system and to then interrupt them with the appropriate NSMs that help guide your nervous system back into *the window of tolerance*. It's not the random application of stress management techniques to temporarily 'make you calm' but the strategic and powerful application of both habit change and polyvagal informed interventions to retrain your nervous system to function as it was meant to.

> **The goal isn't calm. It's nervous system regulation.**

It's All about the Vagus Nerve

In order to sustain life, the body has two complementary systems within your autonomic nervous system: the sympathetic and the parasympathetic. As you read in Chapter 4, the sympathetic has one branch and the parasympathetic has two, the ventral and the dorsal. They are all essential for psychological and physiological balance as well as for survival. Without a dynamic and responsive interplay of the sympathetic and parasympathetic branches of the autonomic nervous system our heart would beat too quickly or too slowly to sustain life, our respiratory system would struggle to maintain the balance between inhalation and exhalation and the release of hormones would become imbalanced, erratic and uncoordinated. For the systems in our body to work in balance and harmony the nervous system needs to be operating in synchrony. As you will recall from the last chapter, this balance can become compromised and we can get stuck in survival mode. The balance we need - to be thriving, resilient and flexible human beings - is achieved via the vagus nerve.

Amongst its many roles, the most crucial role of the vagus nerve is as the chief conductor of the parasympathetic nervous system - thus the chief conductor of the relaxation (and regulation) response. Roughly 80 per cent of its pathways are responsible for transmitting sensory information from the body to the brain. These are called afferent pathways. While the remaining 20 per cent govern motor commands sending messages from the brain to the body. These pathways are called efferent. This nerve serves as the sensory relay system, keeping the brain informed about organ activities, in the skin, the digestive tract, lungs, heart, spleen, liver and kidneys. Because of its branches and innervations, the vagus nerve plays a critical role in various aspects of human communication, from speech and facial expressions to interpersonal connection, operating largely at an interoceptive level - beneath conscious awareness. As you will learn, nearly every practice, system and method of healing you can think of - whether it be breath work, yoga, qigong, acupuncture, massage therapy, gong baths, singing, cold plunging or somatic trauma therapy - largely achieves its therapeutic outcomes via the vagus nerve. In fact, you have already been stimulating and strengthening your vagus nerve with the practices and exercises you've been learning so far.

Synchrony of the Vagus

When the various branches of the vagus nerve are strong, responsive and in synchrony, that's when we feel connected, grounded, present and engaged, and we can recover smoothly from stress. The term *vagal tone* represents the activity and effectiveness of the different branches of the vagus nerve. High vagal tone means a higher capacity for adversity, change and resilience, and a higher capacity for rest, recovery and healing. The higher our vagal tone, the more stress we can take on while staying present and within our window of tolerance. Likewise, a low vagal tone means the opposite: less tolerance for stress, change and resilience while making it more difficult to rest, recover and heal. The lower your vagal tone, the easier it is to become dysregulated.

As it turns out, heart rate variability (HRV), which is now a feature of many fitness trackers, is used by researchers to determine the quality of vagal tone. HRV measures the amount of time (minute fluctuations) between heartbeats during various levels of activity. In simple terms, a healthy heart doesn't beat like a metronome, where each beat is exactly the same distance apart. Instead, it has some irregularity, and those subtle changes in timing are what we call HRV.

The higher your HRV, the stronger your vagal tone is; the lower your HRV, the lower your vagal tone. Why is this? When your heart rate is more variable, it usually means you're in a lower-stress, more relaxed state. On the other hand, low HRV is often linked to stress and health issues. So, HRV can be a useful indicator of your overall well-being and how well your body is handling stress. You now understand that the essence of nervous system dysregulation is that the system gets stuck in extreme sympathetic or dorsal vagal reactions, or see-saws between the two. Therefore, another way to think of vagal tone is as *vagal synchrony*, the degree to which your nervous system can flow in a dynamic and fluid manner between ventral, sympathetic and dorsal states. The ventral, sympathetic and dorsal all have nerves that directly connect to your heart and as they transmit signals, the interplay of these three states is reflected in HRV. When vagal tone (or synchrony) is high, the dynamic yet balanced interplay (signalling) of the three branches is reflected in more fluctuation (variability) in the amount of time between heartbeats. But when tone is low, and instead of dynamic synchrony only one of the vagal states dominates, then there is no interplay of signals, leading to fewer fluctuations and a lower HRV number. The images below show the HRV of a regulated person versus a dysregulated person with PTSD.[3]

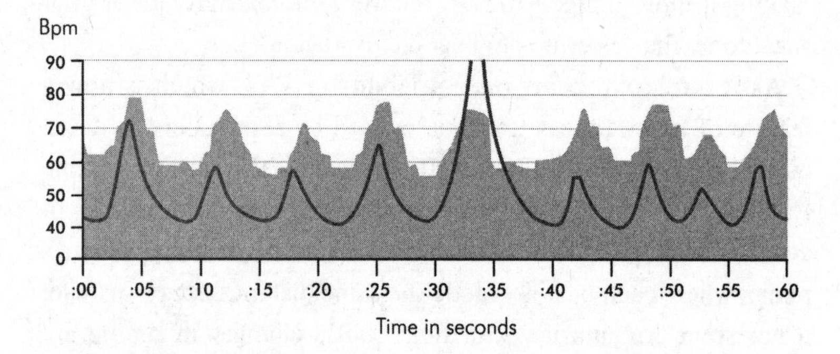

Heart rate variability (HRV) in a well-regulated person. The rising and falling dark lines represent breathing in this case slow and regular inhalations and exhalations. The light grey area shows fluctuations in heart rate. Whenever this individual inhales, his heart rate goes up; during exhalations his heart rate goes down. This pattern of heart rate variability reflects excellent physiological health.

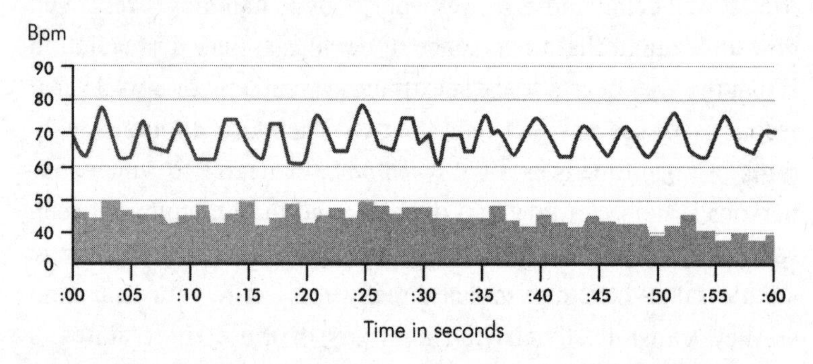

HRV in PTSD. Beathing is rapid and shallow. Heart rate is slow and out of sync with the breath. This is a typical pattern of a shut-down person with chronic PTSD.

Trauma and chronic stress forces our nervous system to occupy survival states like extreme sympathetic or dorsal responses. Over time, due to the nature of the brain, these survival patterns become habituated and our nervous system becomes a one-trick pony (or two tricks, in the case of the seesaw), always defaulting to one of the vagal pathways and incapable of using the others. And, if you recall in the earlier chapters about a trail in the forest, when those circuits aren't being used, it becomes more difficult for us to access and stimulate the ventral vagal branch. More specifically, according to neuroplasticity researchers Jeffrey Kleim and Theresa Jones, 'neural circuits not actively engaged in task performance for an extended period of time begin to degrade'.[4] Without use, the ventral vagal branch loses tone, cannot participate in our regulation and further traps us in survival mode. What a vicious cycle.

Since the dynamic interplay of the three states, especially regular access to the functions of rest, recovery and healing, of the parasympathetic pathways is required for homeostasis, the lack of interplay results in increased allostatic load. Thinking back to the Introduction, a high allostatic load strains your body's ability to maintain basic physiologic functions. At the cellular level, this strain brings about changes that impact on energy production, immunity, hormone signalling and cellular repair, ultimately affecting your overall health. This is how a 'stuck' nervous system makes things worse by contributing to its own cumulative load of stress, which then causes it to intensify its efforts with the same default survival response and maintains the exhausting feedback loop that is the nervous system paradox.

The way out of the paradox is to interrupt the habituated nervous system pattern keeping you stuck with the vagal toning, synchronising and *modifying* power of the NSMs. A toned and healthy vagus response is crucial to recover from stress, injury or illness. In fact, recent research even demonstrates that it is *the* missing link to healing from trauma, anxiety and inflammation, and to alleviating symptoms for mind and body conditions. Even for a physical condition like obesity, animal studies have shown that selective activation within particular vagus nerve branches can suppress the inflammation associated with obesity and reverse metabolic complications.[5] Luckily for you, there will be no missing link. By the time you have read the whole book, and incorporated these practices, you will be the expert of your own vagus nerve.

The Super-Regulating Power of the Vagus Nerve

In order to understand how the NSM exercises are able to assist in *modifying* your nervous system and regulating it, we will dive deeper into the function of the vagus nerve itself.

Imagine you're about to walk into a job interview. As you approach the interview room, your sympathetic nervous system becomes active as it perceives threat and danger. It triggers the release of stress hormones, including cortisol and adrenaline. These hormones prepare your body for action. Your heart rate increases, your muscles tense, and your breathing quickens. This response is essential for alertness and readiness to face a challenging situation like a job interview. You feel a surge in energy and instinctively decide to shake it off before you step inside. So, you take some deep breaths, shake your body, clap your hands together and tell yourself 'Come on, I CAN DO THIS!' and instantly feel more grounded. Your vagus nerve begins to counteract the stress response by releasing the neurotransmitter acetylcholine. This acetylcholine binds to certain receptors and stimulates muscle contractions in the parasympathetic system to calm you down.[6] In addition to acetylcholine, the vagus nerve influences the release of other chemicals, namely prolactin, vasopressin and oxytocin. These substances promote relaxation and social bonding. Prolactin is often associated with reducing stress and promoting a sense of well-being. It helps you maintain composure and clear thinking during the interview. Vasopressin is a hormone that is critical in helping you reduce anxiety and help with social interactions, which is crucial when you're meeting potential employers. And finally, research has documented links between the vagus nerve and oxytocin, even before a baby is born.[7] This helps foster trust and bonding between a baby and mother after birth and promote calm, social interactive behaviour. But in this context, it can help you establish rapport with the interviewers. So, while the sympathetic nervous system initially prepares you for the challenging situation, the vagus nerve, through the parasympathetic response and the release of various calming substances, helps you stay composed, build connections and manage the stress of the job interview. This example illustrates how well engineered our body is to both prepare for and recover from stress and feel calm again. *Modifying* the state of your nervous system through NSMs (in this case through deep breathing, shaking off and clapping) strengthens the recovery response. This is how a healthy and regulated nervous system can instinctively, adequately and helpfully respond to the demands of the environment.

When someone's vagus nerve function is compromised because of trauma or illness, the vagus nerve may not activate the same response. This is because trauma can disrupt the normal functioning of its tone by keeping the body in a heightened state of stress, altering the neurochemical balance, and affecting the brain areas that communicate and regulate it. Let's explore an example of a compromised vagal response. As a result of years of chronic stress, your autonomic nervous system, including the activity of the vagus nerve, becomes dysregulated. Your past traumatic experiences were left unhealed leading to a hypervigilant and overactive sympathetic nervous system and an underactive parasympathetic nervous system. When you face social situations, such as attending social gatherings or job interviews, the overactive sympathetic nervous system dominates your mind and body. The 'fight or flight' response is triggered more intensely and frequently than in other people around you. In this case, the vagus nerve's ability to counteract the stress response is compromised, we are pushed out of our optimal zone and are now operating outside our window of tolerance. The dysregulated vagus nerve may struggle to release sufficient acetylcholine and other calming substances. As a result, you may find it challenging to feel calm even if your cognitive mind is telling your body to chill out. Your experiences in social situations become increasingly anxiety-inducing. Without the parasympathetic influence of the vagus nerve, you may experience heightened heart rate, trembling, sweating and other physical symptoms. Over time, this dysregulation can take its toll, potentially leading to diagnostic labels like social anxiety, depression and chronic pain. Your vagus nerve's inability to regulate the stress response can challenge your ability to form and maintain relationships and perform well in social or professional settings, which will make you feel more isolated and cut off from support, making healing even more challenging.

Typically 'social anxiety' might be treated with anti-anxiety medication but you'll discover – as thousands of our clients have – that simply doing the exercises in this chapter to help you move out of stuck survival states can cause a quantum leap in regulation and healing.

> **The vagus nerve helps you heal, but you heal yourself.**

The A I R approach will regulate your nervous system and help you be conscious of your mind-body-human experience and create a thriving life. NSMs are a powerful biohack because they directly tap into unfelt emotions, nervous system survival states and coping mechanisms through the superhighway of your body - the vagus nerve. And while we aren't saying that NSMs alone will heal you, they will give you the keys to immediate state-shifting and will help you unlock superpower(ful) insights. Just like Dr. Perry Nickelston says, no system ever works alone, never gets injured alone and never heals alone. Functionally stimulating the vagus nerve has immediate immense benefits on all the systems and organs it's connected to. Like most of our clients, we predict that they will become some of your favourite nervous system healing tools. Read Jules' story for a sense of just how powerful these simple exercises can be.

Jules *is a successful painter and for years would spend hours creating art in her studio. One winter, she had a car accident and her right arm, her painting arm, became partially paralysed. Although over time she regained function and movement in her hand and arm, for five subsequent winters Jules would not paint a single thing. She developed depression, anxiety and chronic low back pain, until she was officially diagnosed with fibromyalgia. Jules found our online programme through Jen's Instagram and after a discovery call, enrolled. Her initial message to us read: 'As you know, I am hoping to recover from pain. It all started with a car accident five years ago and it's been a tough time ever since. I am an artist and am hoping to get myself to a place where I can start painting again. I feel stuck and I hope you can help.' That was the last we would hear from her for some time.*

Right from the start it was clear that Jules experienced resistance to using the tools and after some weeks shared that she didn't think the programme was right for her. We offered her two options, to either stay on until module five, or to have a full refund and exit the programme – she opted for the former. We were in our fifth weekly group coaching call when Jules showed up and hesitantly raised her hand. 'I'm a bit scared to speak because as you know I haven't been showing up for the coaching calls and I'm ashamed of it. I also haven't actually completed much of the programme, but I am here now and I don't really have a question, but I'm nervous and I would appreciate any help.' Jen stood up from her desk and began coaching her. She invited Jules to stand up from her chair too, to shake her whole body for about a minute to release the energy of fear that was trapped in her body with nowhere to go. Then, while still standing, she invited her to inhale and then exhale loudly while saying 'Aaaaaaaa'. They repeated this eight times. Then she asked her to swap the 'Aaa' sound for an 'Aaaaaaaooooouuuummmmm' sound and they repeated the breath eight times again. They ran through this entire sequence three times and then sat down. Jen coached her to press her palms against her eyes for 30 seconds. Jules' posture softened and suddenly she began crying until she was sobbing. She continued to sob for some time and the other 30 people on the call started writing messages in the chat.

'I love you Jules.'

'I'm proud of you!!'

'We are in this together Jules!'

'You've got this, I promise.'

'I feel this so deeply, thank you for being vulnerable with us. I deeply appreciate you!'

'You are worthy, we are all here for you.'

'We love you Jules!!!!'

'SO proud of you for being vulnerable and brave!!'

'Let yourself cry, we are here for you. I see you.'

Some students began crying too. Jen and Karden were witnessing a wave of energy in motion through expression and feeling – they were witnessing the language of the body. The feelings in the room began to gently shift and nervous systems began to modify their responses from sympathetic to ventral vagal and connection. After the call, Jules immediately booked a few sessions with Jen and Karden and they began working with her to help her catch up on the first five modules. Here's a letter Jules wrote ten weeks later.

Dear Fibromyalgia
I hated you for coming into my life, turning it upside down and keeping me stuck. I hated you for making me feel so much pain – both emotionally and physically. I hated you for keeping me away from my art. I hated you for taking away years of my life. Countless are the nights I cried myself to sleep wondering – WHY ME, WHY? Now I know why. One night, I was driving my car and slipped on a frozen road and crashed into a ditch. This forced me to stop painting and in the months following, I met my darkest shadow self. She was angry, sad, mean, helpless and terrified. She was also numb and wanted to crawl away in a dark cave and stay there for a thousand years. In those months I started to meet you. At first you would come and go until you decided to stay. You gave me aches on my skin, then in my muscles and in my bones. You gave me even more darkness than I thought possible. I would pretend you didn't exist, but then you became even louder and it was impossible to ignore you. Years went by, five to be precise. I learned a lot about you and about me. Then one day something changed. I read something somewhere about you having a purpose and being in my life for a reason. I came across nervous system work, and it all started making sense. You showed up not to keep me stuck, not to take my life away and not to make it harder – but to show me that I had more healing to do before I could move on to the next stage of my life. At

first, this enraged me. But now – I know I wouldn't be where I am without you. The car accident was the last straw to break the camel's back, then you showed up not to make me sick but to help me heal. Because of you, I healed myself so deeply that I am better than I ever was. Because of you I learned how to speak with my body as if we had our own language. Who knew that pressing into my eyes would help me feel again, that tapping my chest would instantly show me that fear isn't permanent and that voo-ing and aaa-ing would show me that I am capable of changing the response in my nervous system. Thank you for all the lessons and, in hindsight, thank you for all the pain. This is not a farewell, this is a goodbye.
Sincerely
Jules

> **Modifying the responses in our nervous system helps us find the parts of us that have been trapped and shows them a way out.**

How NSMs Work

In 1872 Charles Darwin wrote:

> a sensitive nerve when irritated transmits some influence to the nerve-cell, whence it proceeds; and this transmits its influence, first to the corresponding nerve-cell ... and then upwards and downwards along the cerebro-spinal column to other nerve-cells, to a greater or less extent, according to the strength of the excitement; so that, ultimately, the whole nervous system may be affected. This involuntary transmission of nerve-force may or may not be accompanied by consciousness.[8]

Translated into plain language, Darwin is saying that if you stimulate a nerve, it gets activated and sends electrical-chemical energy to all the other nerves it's networked to. For example, when the specialised nerve cells in your nose called olfactory receptors get stimulated by the smell of freshly baked chocolate chip cookies, they get excited and send electrical-chemical energy to our brain and our stomach, which causes us to salivate and initiate digestive functions.

In a similar way, certain stimuli can activate the vagus nerve and cause it to send electrical-chemical energy throughout your nervous system and to your brain. Because the vagus (wandering) nerve is widely and generously invested into your body, there are lots of ways to stimulate and excite it. As you can see in the illustration on page 124, the ventral branch of the vagus nerve in particular is connected to your face, eyes, ears, mouth, throat, chest, heart and lungs. Actions as simple as massaging, tapping or vibrating (with voice or sound) are sufficient to stimulate and excite the ventral vagus nerve and cause it to send its regulating effects throughout the nervous system. Breath work, yoga, qigong, acupressure and toning derive many of their therapeutic effects by the ways they stimulate the vagus nerve. The NSMs that you will be doing momentarily are all specifically designed to target the vagus nerve and stimulate its regulatory effects.

Moreover, the principle of mutual inhibition states that in a system where there are opposing aspects, like the sympathetic response versus the parasympathetic response, when one is stimulated, the other is inhibited. For example, if you're in a sympathetic dominant response, NSMs that stimulate the vagus nerve can not only increase the regulatory activity of the vagus, they can *decrease* that activity in the sympathetic branches of the nervous system. Conversely, if you are in a dorsal dominant state, NSMs that stimulate the sympathetic and ventral aspects of the nervous system will upregulate you and inhibit the shutdown effects of the dorsal vagus. And the best part is that these practices don't require any participation of the mind, they are entirely physical and all you have to do is do them to get the results.

So without further ado, let's start *modifying*!

Let's summarise what you know now about modifying for interruption in the body:

- NSMs are designed to help you move your nervous system out of the patterns it has become stuck in and return it to adaptability and flexibility.
- These patterns are survival 'habits'. It is through the interruption of these that the nervous system can rewire itself for safety habits.
- The goal is to move in and out of states, returning to our window of tolerance with ease. The goal isn't calm. It's nervous system regulation.
- The vagus nerve is the most important nerve in our body that activates our super-healing powers.
- Vagal synchrony represents nervous system regulation. And through HRV we can determine whether our nervous system is in synchrony or not.

NSMs help us achieve vagal synchrony and modify the dysregulated responses, transforming them into regulating ones.

Practices 5
Interruption: Modifying

Learn to modify the default survival responses of your nervous system

In these practices you will learn NSMs that tone your vagus nerve, discharge repressed survival energy and mobilise yourself out of shutdown.

Why Do It?

Modify techniques help you interrupt unhelpful body patterns through leveraging the healing power of the vagus nerve. By repeating these, your nervous system will be receiving direct input that will functionally trigger a soothing and regulating response and help move toward a ventral vagal response.

When to Do It

When you identify the survival default state(s) you are stuck in, use NSMs to shift you into a ventral vagal state.

Tips before you begin

- Experiment! We are complex creatures and everyone responds differently to the NSMs. Try the various techniques multiple times and in response to multiple different types of triggers and note which ones work best for you.
- If something doesn't work for you, you don't have to do it. That's normal.
- At times, we can be so activated that identifying our survival state is very difficult; in that case, it's perfectly OK to just start doing your favourite NSMs. As you begin to regulate, your Observing Self will be able to come online and you'll be able to listen and translate.

Practice 1: Regulate the Ventral

Use the following practices either independently from one another or in a sequence when you feel overwhelmed, stuck and getting mixed messages and signals from your body. Over time, the cumulative effect of practising these will tone your vagus nerve and help strengthen your nervous system's responses for regulation and healing.

Box Breathing

1. Find a comfortable place to sit or lie in.
2. Notice and describe how you feel before you begin the practice.
3. Place one hand on your heart, and the other on your belly to help you connect with your body through your breath.

4. Breathe in the following pattern for up to 5 minutes:
    a. Inhale for four counts.
    b. Hold your breath at the top of your inhale for four counts.
    c. Exhale for four counts.
    d. Hold your breath at the end of your exhale.
5. Check in with yourself and notice already how different you feel compared to when you started.

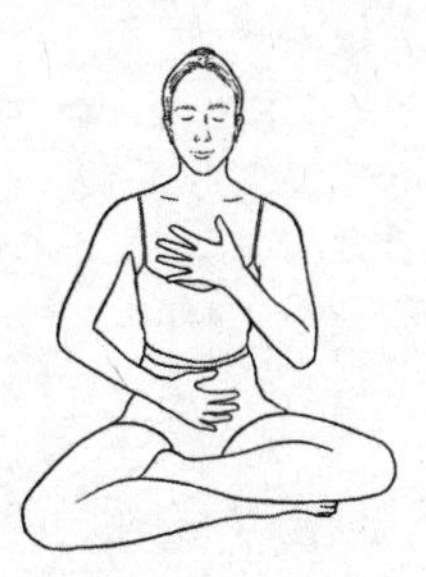

Eye Relaxation

1. Notice and describe how you feel before you begin the practice.
2. Bring your palms to your eyes and gently hover over them for a few moments without making any direct contact.

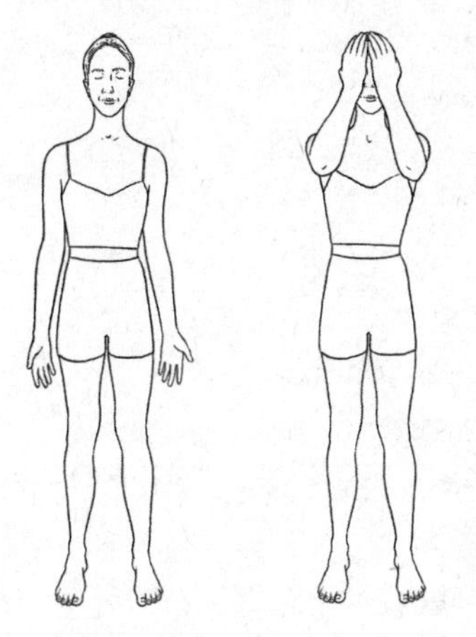

3. Feel the warmth.
4. Take a deep breath in through your nose and slowly exhale through your mouth.
5. Then gently press your palms into your eyes until you see a 'light show' behind your eyelids.

6. Check in with yourself and notice already how different you feel compared to when you started.

Ooo Aaa Eee Tapping

1. Find a place where you will be undisturbed and comfortable making funny sounds.
2. Notice and describe how you feel before you begin the practice.
3. Two things will happen at the same time:

a. Take a deep breath in through your nose and then in one exhale only consecutively say: Ooo – Aaa – Eee. Stay on each vowel for up to 3 or 4 seconds.
b. Tap your chest with your hand.

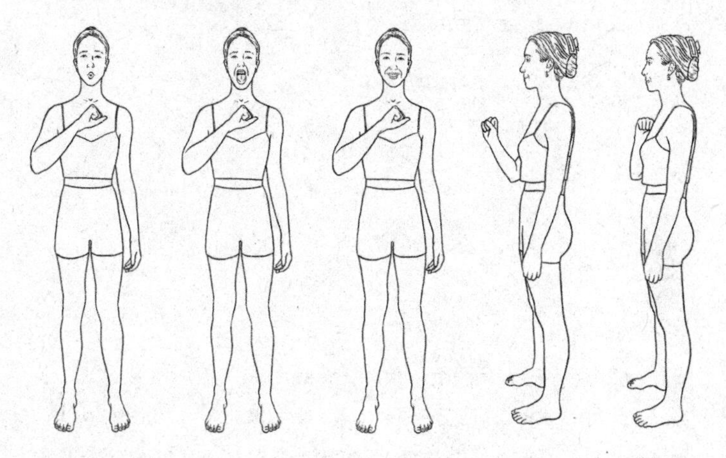

4. Repeat for up to 5 minutes.
5. Notice and describe how you feel now that you've completed the practice. What's shifted?

Head Tilt

1. Find a comfortable place to sit in.

2. Notice and describe how you feel before you begin the practice.
3. Inhale and loudly sigh three to five times.
4. Bring your right ear towards your right shoulder, without moving your shoulder.

5. Look up to the far top left corner.
6. Stay here for up to 30 seconds (a little bit of dizziness is normal – if you feel dizzy, decrease the time spent here to 5 or 10 seconds).
7. Pause for a moment. Maybe walk around and move your body.
8. Then, inhale and loudly sigh three to five times.

9. Bring your left ear towards your left shoulder, without moving your shoulder.
10. Look up to the far top right corner.

11. Stay here for up to 30 seconds (a little bit of dizziness is normal – if you feel dizzy, decrease the time spent here to 5 or 10 seconds).
12. Notice and describe how you feel now that you've completed the practice. What's shifted?

Practice 2: Downregulate the Sympathetic

Use the following practices either independently from one another or in a sequence when you feel angry, anxious and activated.

Stomp and Sigh

1. Find a place where you can comfortably be barefoot (ideally in nature on grass, earth, sand, moss, mud ...).

2. Notice and describe how you feel before you begin the practice.
3. Barefoot, alternating, stomp one foot at a time on the ground for up to 2 or 3 minutes.

4. Then inhale normally through your nose and before you exhale, quickly inhale the remaining air in that same breath (double inhale).
5. Exhale slowly through your mouth.
6. Repeat three to five times.
7. Notice and describe how different you feel now that you've completed the practice. What's shifted?

Shake and Twist

1. Find a safe place you feel comfortable moving around in.
2. Notice and describe how you feel before you begin the practice.
3. As if trying to drip water off your hands, arms, legs and feet and finally your whole body, shake yourself for up to 2 or 3 minutes.

4. Then stand and open your feet wider than your shoulders.
5. Sway your arms around your body allowing your head and torso to twist all the way to one side and your hands and arms to wrap around your body as you twist.

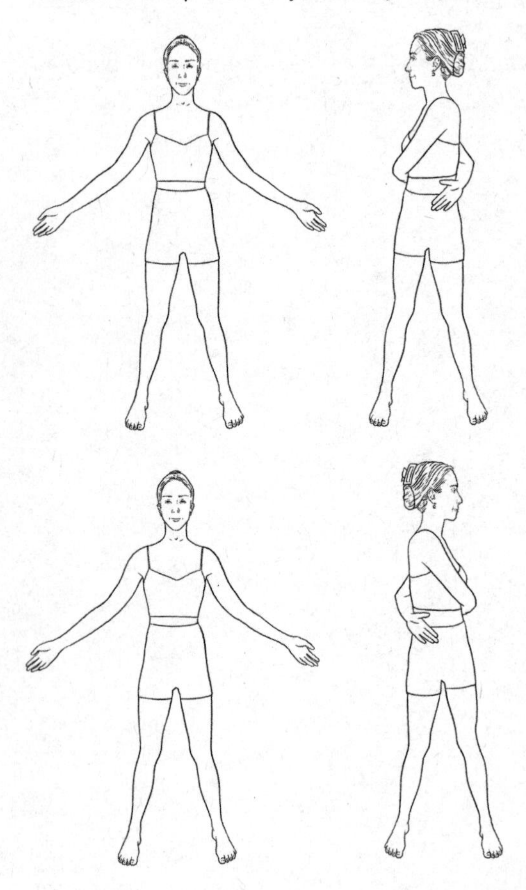

6. Repeat on the other side.
7. Do this in one single movement and repeat for up to 3 to 5 minutes.
8. Notice and describe how you feel now that you've completed the practice. What's shifted?

Contract and Extend

1. Find a comfortable surface to lie on.
2. Notice and describe how you feel before you begin the practice.
3. Extend your arms by your sides and your legs in front of you and relax.
4. For 10 seconds contract your whole body in a forward shape and purse your lips.

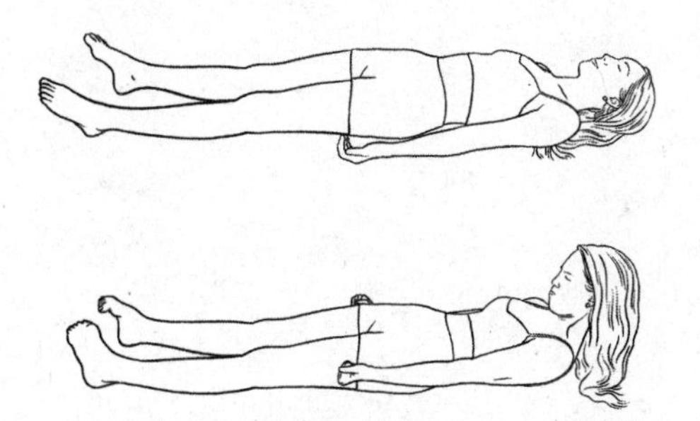

5. And relax.
6. For 10 seconds, extend your whole body in an arching shape and open your mouth.

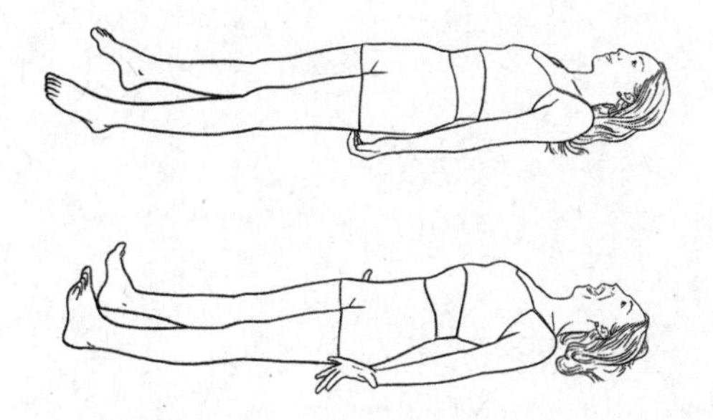

7. And relax.
8. Repeat this contrast for up to 2 or 3 minutes.

9. Notice and describe how different you feel now that you've completed the practice.

Breathe and Fold

1. Find a comfortable place to sit or stand in.

2. Notice and describe how you feel before you begin the practice.
3. Inhale through your nose and then exhale through your mouth as if breathing out through a straw, emptying your lungs.

4. As you do this, fold your upper body forward.

5. Take five of these breaths.
6. Take a break in between breaths.
7. Notice and describe how different you feel now that you've completed the practice.

Practice 3: Upregulate the Dorsal

Use the following practices either independently from one another or in a sequence for when you feel sad, depressed, shut down and dissociated.

Find Rhythm

1. Find a safe place you feel comfortable moving around in.
2. Notice and describe how you feel before you begin the practice.
3. Stand and spread your feet, hip width apart.

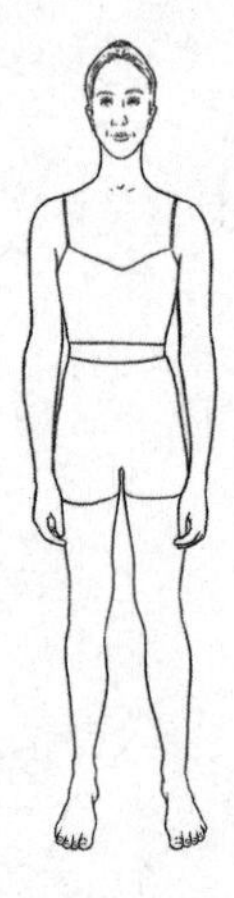

4. Begin swaying however your body naturally sways.

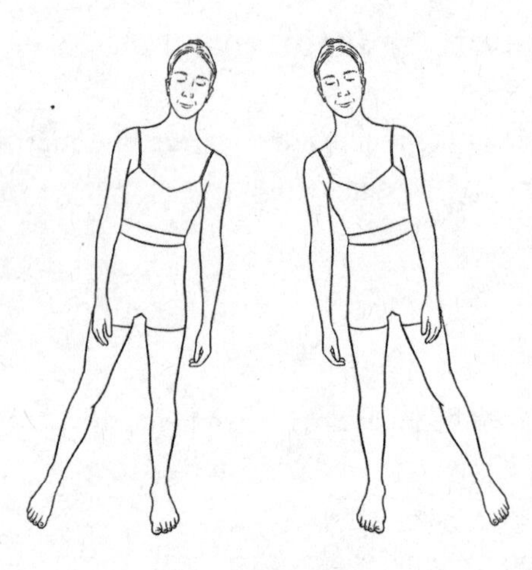

5. Sway for up to 5 minutes.
6. Notice and describe how you feel now that you've completed the practice. What's shifted?

Find Flow

1. Find a safe place you feel comfortable moving around in.
2. Notice and describe how you feel before you begin the practice.
3. Stand and spread your feet, wider than your shoulders.

4. Put your hands on your hips.
5. Begin gently moving your hips in a figure of 8.

6. Do this for up to 5 minutes.
7. Notice and describe how you feel now that you've completed the practice. What's shifted?

Regulating Ears

1. Find a comfortable place to sit in.
2. Notice and describe how you feel before you begin the practice.
3. Begin massaging your ear in the areas pointed out in the illustration below. Do this for as long as it feels soothing.
4. Deeply sigh.
5. Check in with yourself and notice already how different you feel compared to when you started.

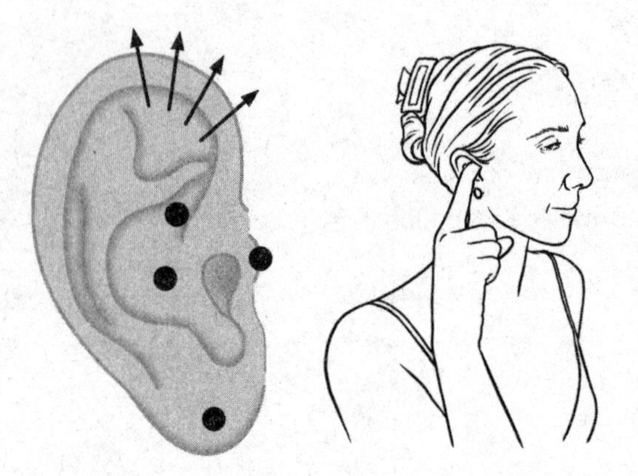

Regulating Eyes

1. Find a comfortable place to sit or lie in.

2. Notice and describe how you feel before you begin the practice.
3. Look to your far right, while facing forward without moving your head.

4. Inhale and exhale while sticking your tongue all the way out.

5. Repeat this on the other side.

6. Do this one to two times per side.
7. Notice and describe how different you feel now that you've completed the practice. What's shifted?

Make Modifying a Part of Your Life

The goal is to make a habit of using NSMs to regulate your nervous system and help you become more embodied. To do this, we invite you to:

1. Take a moment each day to check in on your vagal system. Notice what energy you sense and truly become familiar with how your body best responds and shifts from survival mode to a helpful mode.

2. When you recognise what feels good, notice whether you can introduce the same technique to your lifestyle and self-care routines.
3. Try practising yoga and qigong or kick-boxing and karate, going for a massage or self-massaging the parts of you that feel soothing, spending more time in nature, learn new breath work and tapping techniques and so on and so forth.

Notes:

....

....

....

....

....

....

Chapter 6: Redesign

Settling

your nervous system to feel at home in your body

It feels like coming home, after being gone too long.

Unknown

Throughout this book we have looked at the consequences that so many people are living with due to a nervous system that is chronically stuck in a state of survival. Those consequences include everything from anxiety and depression to a variety of chronic illnesses. But underlying all of these consequences is a deeper and more essential loss. The loss of your birthright to feel at home in your own body. Feeling at home in your body is the sense of safety we often talk about in this book. And having a felt sense of safety literally means experiencing comfort, ease and contentment within yourself – like curling up with your favourite blanket in your favourite chair in front of a warm fire with no place to go, nothing to do and no distractions. This is the feeling of being in the ventral vagal mode of your nervous system where you feel calm and comfortable, open-hearted, connected to others, mindful and present – or, in one word, settled.

As long as we are unable to access a sense of safety in our body, it's impossible for us to arrive at home in ourselves and our nervous system will struggle to stay within the window of tolerance. Fortunately, now that you are learning to listen and speak the language of the body, you can begin to guide the conversation toward a felt sense of safety and explore how good it can feel to be at home in your body. With the understanding you now have from polyvagal theory and your ability to move your body in and out of different states with nervous system modifiers, our goal in this chapter is to use *settling*, the redesign techniques of A I R in the body, to support you in arriving home and making safety, ease and connection your default setting.

It sounds simple, doesn't it? Well, there is a catch. Initially, nervous systems that are in a state of dysregulation often have difficulty inhabiting states of safety, ease and connection. As we take a closer look at the way the brain has evolved, you'll understand why the nervous system may initially resist your efforts to change it and create new and useful healing states. Resistance is rooted in the very nature of how evolution has wired our brains to function and can be explained through the principles of familiarity and negativity bias. These principles are the main obstacles in making a quick shift from our survival default states to thriving ones.

By learning how these principles influence you and that they are in fact a normal part of the process of healing your nervous system, you'll be able to apply the *settling* practices more effectively to make thriving states your new normal. Ultimately, the goal is for your nervous system's default mode to be safety and regulation rather than high alertness and survival. The *settling* practices at the end of this chapter will help redesign your nervous system to be more comfortable with the changes required to regulate it and help you heal.

Familiarity

The familiarity principle is simple. The brain always prefers what it knows to what it doesn't. Don't we all? In fact, psychological researchers have labelled this phenomenon the *mere-exposure effect*, which states that people will start liking something simply because they have been exposed to it many times and prefer that item over something new. Karden remembers a vivid example of this from his early twenties. After growing up in suburban New Jersey and living a mainstream life of eating fast food and working at the mall, he decided that he wanted to learn how to become a massage therapist and went to massage school in Ithaca, New York. In the midst of his studies he decided to start eating healthily and a friend told him that the place to shop for healthy food was the local food co-op. Until that point in his life, he had only ever been in big conventional grocery stores, so the Green Star Food Co-Op in Ithaca was like another planet to Karden. The smells were different. The layout was different. The products were different. The people were different. And he remembers being distinctly uncomfortable in the store and walked out after about 20 minutes without buying anything and swearing to himself that he wouldn't be going back. As it turns out, he did go back, and through repeated exposure now the opposite is true: co-ops are where he now feels at home and he feels less comfortable in conventional supermarkets.

Undoubtedly, you have experienced something similar. And when it comes to the language and experience of our body, if we are used to living in dysregulated survival states, they are familiar and preferred by our brains. The regulated thriving states that you are learning to inhabit, being unfamiliar, will often be initially rejected by your brain. This is a natural part of the process. There is nothing wrong with you if you are having difficulty staying in a more helpful window of tolerance because your brain and body need time, patience and practice to become familiar with what you are guiding it toward. As the familiarity principle implies, the more you can begin to practise *settling* techniques in your daily life and expose yourself to useful states, the sooner your brain will learn that they are familiar and trustworthy.

Negativity Bias

The negativity bias states that even when of equal intensity, things of a negative nature (e.g. unpleasant thoughts, emotions or experiences) have a greater effect on one's psychology and physiology than neutral or positive things.[1] In other words, our mind and body will generally look for and respond more intensely to negative experiences than to positive ones. Our brains evolved to do this because when employed in a measured way, negativity bias has aided our survival. After all, no one has ever died from forgetting to get ice cream from their favourite ice-cream shop, but someone has died by mistaking a poisonous snake for a harmless stick and getting bitten. It was important that our ancestors remembered where the poisonous berries were. Since negative things always grab our attention more than positive things, the negativity bias strongly influences the media as well. When you watch the news you may have wondered why everything seems so negative. One study examined reactions to video news from 17 different countries and found that globally, humans are more physiologically activated by negative news stories than positive ones.[2]

Unfortunately, when we are chronically stuck in survival states or dealing with unhealed trauma, our negativity bias gets amplified and the nervous system is busy continuously scanning for threats.

This often shows up in the form of overthinking, worrying and catastrophising, and is felt in the body through looking for and scanning for unpleasant sensations or symptoms. It also results in overreacting to perceived threats. In doing so, the negativity bias and the familiarity principle form a feedback loop that reinforces a pattern of looking for danger, which unfortunately makes feeling unsafe in one's body a familiar feeling, and therefore the nervous system's preference. Although this is inconvenient, as with the familiarity principle above, it is normal and, with *settling* practices, changeable. A wonderful instance of this was our client Abby.

Abby *was a 'Type A' high performer who loved her job but was headed toward total burnout. She told us that her biggest challenge was that she was always on the lookout for fires to put out and simply couldn't unplug from work. This gave her headaches, back pain, fatigue, exhaustion, emotional overwhelm and blurry vision. Even when she was on vacation with no one requiring her presence at work, she would check her emails incessantly. No matter how much she told herself not to or tried to resist the urge to be hyper-vigilant, she was powerless against the impulse of her nervous system. But as Abby began to work with us and learn the language of her body, she was relieved to discover that there was nothing 'wrong' with her – her nervous system was plainly doing what it was best designed to do. In this case, it was defaulting to the familiar, high-stressed states that served the purpose of keeping her safe. While achieving is associated with positive self-fulfilment, with past unresolved traumatic experiences high-achieving behaviours can lead to burnout because of the patterns driving these. We coached Abby to investigate her belief systems stemming from her unhealed trauma. 'As long as I achieve, I feel worthy. When I stop achieving I feel afraid and worthless.'*

We helped her learn what these felt like in her body in order to recognise the evident link between symptoms and patterns in her mind–body–human aspects. Much to her surprise, simply bringing up her beliefs about herself caused a lot of anxiety and an instant worsening of the aforementioned symptoms. As we delved further into A I R, there was one step that created the first real shift – this was settling. We taught her to use the butterfly hug, the containment hug and the self-soothing exercise after tuning into BASE-B and after some days, she began feeling, maybe for the very first time, like she was finally arriving home. Nowhere to be, no email to reply to, no hustle, no pressure, no headache, no back pain. A worthy and whole Abby.

'This may sound strange, but I don't think I ever felt like I had a body – with feelings, sensations and "messages". I've just been living in the upbeat, fast and overwhelming space of my mind my whole life. In the career I'm in, there are a lot of high-achieving performers and I always wondered why everything was such a struggle for me. I often shamed myself for feeling tired or exhausted, even hated myself for it. Now I know. I was completely disconnected from the feelings in my body shouting to be kind, instead of mean. Love myself instead of hate myself. I didn't know that there was such a thing as moving at the speed of love. Now I do. I feel settled, at last.'

Abby was able to learn how to maintain states of ease and joy, calm and connectedness. Abby told us she doesn't fear burnout anymore because she knows how to turn off 'the burn' in her own nervous system and feel safe. A critical part of the settling practices is to redesign the search for something negative to respond to (and therefore dysregulate you) into spending more and more time in a state of relaxed awareness.

> **You may be a 'Type A' or you may be disconnected from the feelings in your body.**

Hopeful reframe: Knowledge is power. Now that you know about familiarity and the negativity bias, we hope that you can now see that there is nothing wrong with you if you have had trouble relaxing, feeling at ease or experiencing peace. It's not a fault of your character, intellect or will. It's the result of the particular ways our brain is wired and how that wiring interacts with stress and trauma that keeps you stuck. But now that you understand the causes, you can use the same neurological processes that got you stuck to get you unstuck. And as you can observe your resistance with knowledge, understanding and compassion, you will be able to ride the waves and find your way to calmer seas.

Settling into the Body via the Vagus Nerve

In Chapter 2, we shared Bessel van der Kolk's insight that 'one of the clearest lessons from contemporary neuroscience is that our sense of ourselves is anchored in a vital connection with our bodies'. Despite the obstacles of habit change described above, we are going to share with you simple and effective ways to establish the vital connection you have with your body and learn how to feel wonderfully at home inside it. You may recall that 80 per cent of the vagus nerve is afferent – meaning that it sends information from the body to the brain. Therefore, when our vagal tone is low our brain receives very little information about what's going on in our body. Moreover, when our vagal tone is low and we are stuck in an extreme sympathetic response or dorsal shutdown, we lose access to the ventral vagal state, which is a state that actually allows us to feel settled in our bodies.

After we use *modifying* techniques to move us out of whatever vagal state we habitually become stuck in, *settling* techniques are dedicated to toning the ventral branches of the vagus nerve so that we can not only become unstuck, but we can actually be in a safe, present and pleasurable experience inside ourselves. By learning how this feels in your body and cultivating sustained access to it, your nervous systems will know how to come back to the ventral state after being triggered and return to homeostasis rather than allostasis. Remember, regulation is not that you never get triggered, it's about being able to return to the optimal zone of your window of tolerance after you've been triggered in a reasonable amount of time.

As with the *modifying* techniques, *settling* techniques work because they directly stimulate physical parts of the vagus nerve. When the nerve is touched or vibrated, it becomes 'excited' and starts sending signals via its afferent pathways to the brain. Since the nature of the stimulation is positive, the vagus communicates this to the brain and we then become conscious of the positive feelings we're experiencing in the body. As we become aware of this novel and positive experience in our body, it's as if our brain 'wakes up' to the fact that this is a possibility for itself - meaning that it has a better alternative to extreme sympathetic activation, dorsal shutdown or the seesaw.

Stephen Porges, Peter Levine, Lucina Artigas, Deb Dana and other pioneering clinicians have developed a host of vagal toning exercises that are simple and effective. These NSMs serve two functions. The first is that they can be used on a day-to-day basis as a go-to option (like the *modifying* techniques) for regulating yourself when you feel activated. The second is that the regular practice of them is cumulative. In the same way that regular exercise leads to the increased strength, endurance and tone of your muscles, regular practice of the *settling* NSMs leads to increased capacity of your ventral vagal system, which expands your window of tolerance and ability to stay in your optimal zone.

The Vooo Protocol

Peter Levine created an exercise called the 'vooo' sound to help his clients feel safe in their bodies. It involves making a deep, low-pitched vocalisation that vibrates in the chest and abdominal area: 'it opens up your chest (heart and lungs), mouth and throat, pleasurably stimulating the many serpentine branches of the vagus nerve'.[3] Making the 'vooo' sound engages the body in a focused and intentional way and allows you to actively participate in the autonomic responses of your body. It brings attention to sensations, promoting a sense of grounding and body awareness. The low-frequency vibration produced by the 'vooo' sound can have a calming effect on the nervous system and helps to release tension and stress that may be held in the body's tissues, which contributes to a feeling of relaxation and safety. As you will find out in the practice section, it has an immediate soothing effect on your whole self. This can be particularly helpful for individuals who have experienced trauma, as it helps them reconnect with their bodies in a safe and soothing manner. It has been used with hundreds of thousands of clients all around the world with great success.

The Butterfly Hug

The butterfly hug was first employed by Lucina Artigas in 1998 as she was working with traumatised hurricane survivors in Acapulco. The goal of the exercise was to help the survivors access a felt sense of safety in their bodies. It involves interlacing your thumbs, crossing your hands across your breastbone and collarbones, and initiating a rhythmic alternating tapping. This action stimulates the substantial vagus branches that inhabit the chest and throat. Its rhythmic nature helps one feel soothed and calm when in a state of fear, freeze or shutdown. In addition, the bilateral alternating tapping has somewhat similar effects to eye movement desensitisation and reprocessing (EMDR) therapy in helping the brain process traumatic events by decreasing physiological hyper-arousal.[4]

The Containment Hug

The last technique leverages an aspect of the vagus nerve that we will touch on in the next chapter, called the social engagement system. The containment hug was developed by Peter Levine. It allows us to self-simulate the act of being held in a protective and supportive manner by another person. By stimulating afferent pathways of the vagus nerve that send signals of co-regulation and social support to the brain, the containment hug is a potent technique for allowing us to feel safe and maintain a ventral vagal state.[5]

Lawrence *sent Jen a message on social media one day ...*

> 'Hi Jen, I saw your post about anxiety and it really resonated with me. I lost my mother when I was 9 years old and although I am now 27, I still experience a lot of grief. I have been in therapy for years (since I was 11), and have worked extensively on my grief but

somehow I still don't think I've processed her death. I'm writing because what you wrote about the mind and body speaking two different languages really hit home and I think I need some help integrating my body into all the healing I've done. I know my anxiety stems from my grief, and I have worked on it in therapy so much, but it still haunts me, it's still loud and so I'm ready to try your work.'

We started working with Lawrence and noticed that he had a high level of awareness in his mind, knew a lot of facts about grief and anxiety, but couldn't seem to put into practice what he knew. There was a clear mind–body disconnect. So we began to work with Lawrence's experience of his body. At first he found it quite triggering to feel into BASE-B, but with gentle guidance over a few weeks, he was able to tune into his breath, actions, sensations and emotions with relative fluency. Lawrence was stuck in a freeze state, almost as if his nervous system had frozen 18 years ago. Anxiety was a way for his nervous system to try and manage the overwhelming feelings (that he wasn't feeling) in his body. We began learning different types of NSMs to help shift his freeze response and move the built-up energy so that he could access the beginning of what would later become an open and toned ventral vagal state. Lawrence loved NSMs and practised them every day giving him an outlet to finally start mobilising his inner world. Then we taught him the settling exercises. The first time he did the full sequence of butterfly hug, into vooo sound, into containment hug, this happened:

'Do you have time for an emergency session tomorrow? I need to know what happened in my body and I would love you to help me understand!!!!

PS: it's nothing bad – in fact I think something miraculous happened.'

Three days later Jen had a session with Lawrence. 'I did the sequence like you told me, and I had the most unexpected response. I don't know why but I was expecting to feel even more sadness and grief but I feel ... joy. I feel my body. I feel the longing for connection. I feel the longing for love. I feel an expansion of my heart – it's like maybe I'm ready to feel love again. I've been so afraid of love because of the pain that came with losing the greatest love of my life, my mom.' Lawrence began arriving home in his body and could finally feel a bit of ease. His nervous system had moved out of being stuck and suddenly all that frozenness transformed into warmth. Lawrence found that shortly afterwards his anxiety significantly improved leaving him with a newfound happiness for life.

Making Settling Your Way of Life

Settling provides a comforting embrace, helping you connect with your body and experience a sense of belonging within yourself, like rediscovering a long-lost home. It can help you feel a profound sense of relief, like an exhale that extends to the bottom of an ocean. You are learning to feel safe. You are learning to feel. You are arriving and settling home. At their most basic, settling practices are about increasing the moments in which you feel soothed and extending the length of each of those moments so you feel soothed more often and for longer. Because of the familiarity principle and negativity bias, settling practices work best when applied with consistency and in small doses at the beginning. The nervous system tends to resist more strongly when it perceives that its default patterns are being redesigned abruptly. In practice, the experience of settling can be a lot like a stone skipping on water. At first, with the stone moving so fast, it skips off the surface of the water, but with each skip it loses momentum and eventually it lands and sinks through. Similarly, the momentum of dysregulated states takes time to land. As you practise, it is normal and appropriate for you to feel your nervous system resisting and literally bouncing off your efforts to cultivate ease in the body. As you remain aware, patient and compassionate towards these rebounds and gently invite yourself back into the practice, the bouncing will become less and you will feel your mind and body settle into the states of safety and ease. One of our clients, Debra, wrote a letter to her past self that describes the journey towards settling:

> I want to tell you that things can change and they will change. They don't become magically easy overnight, but you will suddenly start to see a path where there was only a black hole. You will start to slowly allow layers that have been protecting you to soften and feel underneath them many sensations, emotions and turbulences, but you will find supportive ways to feel them and they will get less intense. By allowing your emotions to flow through your body, you will also feel all the good emotions, in a way that makes your skin tingle and your insides warm. You will let go of trying to numb yourself and make room for new and wonderful sensations. You will start to move your body again with joy and enjoy being in it again for the first time in a very long time.

Beyond the primary *settling* practices that you will learn at the end of this chapter, we encourage our clients to embrace a lifestyle that prioritises strengthening this super-regulation highway as a cornerstone of their self-care routines. Here are several proven approaches and examples that, when integrated, can make a significant difference in settling:

- **Deep Breathing Exercises:** Slow, deep breathing activates the vagus nerve and stimulates the parasympathetic nervous system. Techniques like diaphragmatic breathing, box breathing and the 4–7–8 breath can be effective.
- **Meditation and Mindfulness:** Practices like meditation and mindfulness reduce stress and improve vagal tone, enhancing the vagus nerve's function.
- **Yoga and Qigong:** Yoga and qigong combine physical postures, breath control, visualisation and meditation to calm the nervous system and stimulate the vagus nerve.
- **Progressive Muscle Relaxation:** This technique involves tensing and then relaxing muscle groups, which can promote relaxation and help balance the autonomic nervous system.
- **Biofeedback:** Biofeedback tools such as wearable HRV (heart rate variability) devices provide real-time data on physiological functions, allowing individuals to learn how to monitor heart rate and reduce stress.
- **Cold Exposure:** Cold showers or ice baths can stimulate the vagus nerve and increase its tone.
- **Physical Exercise:** Regular physical activity helps regulate the nervous system, reduces stress and promotes overall well-being.
- **Nutrition:** Consuming a balanced diet rich in nutrients like omega-3 fatty acids (found in fish and flaxseeds), magnesium (found in nuts and leafy greens) and B vitamins can support nervous system health.
- **Social Connection:** Positive social interactions and maintaining meaningful relationships can stimulate the vagus nerve and promote emotional well-being.

- **Adequate Sleep:** Quality sleep is crucial for nervous system regulation, as it allows the body to recover and repair.
- **Grounding in Nature:** Spending time in natural settings, also known as 'earthing' or 'grounding', can have a profoundly calming and grounding effect on the nervous system. Walking barefoot on natural surfaces, sitting or lying on the ground, and immersing oneself in natural environments can help reduce stress and promote vagal tone.
- **Reducing Stressors:** Identifying and managing stressors in your life, whether through time management, mindfulness or lifestyle changes, can improve nervous system regulation.
- **Herbal Supplements:** Some herbal supplements, such as ashwagandha and holy basil, have been suggested to help with stress management and nervous system regulation. However, it's essential to consult with a healthcare professional before using any supplements.

Keep in mind that what works best can vary from person to person, and a combination of these may provide benefits when added to your lifestyle choices.

Let's take a moment to summarise the knowledge on settling the messages from your body for Redesign:

- As long as we are unable to access a sense of safety in our bodies, it will be difficult to calm our nervous systems and stay within our window of tolerance.
- Your nervous system may initially resist your efforts to change it and create new and useful healing states; the brain prefers what it knows to what it doesn't.
- When we are chronically stuck in survival states or dealing with unhealed trauma, our negativity bias gets amplified and the nervous system scans continuously for threats.
- Settling techniques help tone the ventral branches of the vagus nerve so that we can become unstuck and feel safe, present and calm inside ourselves.

Practices 6

Redesign: Settling

Teach your nervous system to settle and arrive home in your body

In these practices you will learn how to settle your nervous system through trauma-treatment-based exercises that help your body become more accustomed to a felt sense of safety.

Why Do It?

The settling techniques will help you redesign your nervous system more permanently so that your body can start feeling regulated autonomously and automatically.

When to Do It

When you feel unsafe and disconnected to your body or you feel yourself being drawn into old unhelpful protective patterns, like hypervigilance or overthinking, for example.

Tips before you begin

- The following exercises can either be done independently or in sequence.
- If your nervous system struggles with safety and soothing, that's OK. In the following exercises you will be gently guided to functionally stimulate the vagus nerve and the parasympathetic nervous system, and you will find yourself settling into the exercises.
- Resistance is normal. Be gentle with yourself and trust the process and your body's healing wisdom.

Practice 1: Soothe Your Body

Butterfly Hug

1. Find a comfortable place to sit or stand in.
2. Notice and describe how you feel before you begin the practice.
3. Gently tune into the experience of your body and notice without judgement.
4. Interlace your thumbs and bring your hands to your chest.

5. Rhythmically tap your chest, alternating hands for three to five minutes.
6. Take a deep breath in and notice how soothing it feels to give your nervous system a rhythm to follow.
7. Check in with yourself and notice how different you already feel.

Vooo Sound

1. Find a comfortable place to sit in.
2. Notice and describe how you feel before you begin the practice.
3. Gently tune into the experience of your body and notice without judgement.

4. Inhale deeply through your nose and exhale while saying the vooo sound out loud.
5. Repeat for 3 to 5 minutes.

Self-Containment Hug

1. Find a comfortable place to sit in.
2. Gently tune into the experience of your body and notice without judgement.

3. Hug yourself with one hand on your shoulder and the other under your armpit.

4. Take a long deep breath followed by a long exhalation. Repeat three times.

5. Swap your hand and arm position.
6. Stay here for a few minutes and notice what it's like for your nervous system to be hugged.

Practice 2: Soothe Yourself

1. Find a quiet place to sit.
2. Take a deep breath.
3. Tune into BASE-B and notice what comes up for you. Notice how much more skilled you already are compared to when you first started. Give yourself a hug, and acknowledge how far you've already come.
4. Then, imagine being in your favourite place. Imagine wearing your most comfortable clothes. Notice how good and soothing it feels in your body. Do this for 2–3 minutes.
5. Think about the connected, compassionate and sage-like qualities of your Observing Self and direct them at yourself. Notice how good and soothing it feels in your body. Do this for 2–3 minutes.
6. Take a moment to be here and feel all of this positive feedback moving up and down your nervous system.
7. Now invite a comforting presence to sit next to you. Perhaps a friend, a loving partner, a pet or an imagined

guardian. Notice in your body what it's like to feel their warm and loving presence. Allow this feeling to spread throughout your whole body until it moves outside you, filling the space around you.

8. Stay here as long as you need, feeling the warmth and soothing that already lies within you. You have everything you need to feel safe and soothed.

Make Settling a Part of Your Life

The goal is to make ventral vagal states your nervous system's preferred default state. To do this, we invite you to:

1. Remember that you are learning how to be the leader of your nervous system and you have the knowledge and skills to guide it towards regulation.
2. The nervous system is governed by the laws of neuroplasticity and therefore the more you practise settling, the sooner it will become your default state.
3. No matter what A I R technique you are employing, apply 2–3 minutes of settling into the better state you've generated at the end of every practice.

Notes:

Part III
Human

I am not a mechanism, an assembly of various sections. And it is not because the mechanism is working wrongly, that I am ill. I am ill because of wounds to the soul, to the deep emotional self ...

D.H. Lawrence

Learning to speak the secret language of the body is not just about feeling safe and well regulated, it's about teaching ourselves to engage fully and powerfully with the wonders and challenges of the most personal parts of us.

The final realm of your nervous system contains the precious and vulnerable emotions that are at the very heart of what makes us human. Becoming aware of these is the key to repairing your seemingly unhealable wounds.

You will learn how to become aware of the powerful influence your developmental social and emotional needs have on the conversations happening in your body and how to speak the language of attuning, tending and bonding to meet those needs.

By healing the developmental dimensions of your nervous system, including meeting your inner child and discovering their needs, you'll be able to savour the aspect of life that makes us whole and fills us with the most joy - deep embodied connection to ourselves and our fellow humans.

In this part, you will learn the A I R practices of your humanness.

Attuning

An **Awareness** technique that allows you to tune into and notice your unmet developmental social-emotional needs that contribute to perceptions of lack of safety and nervous system dysregulation.

Tending

An **Interruption** technique that equips you with the tools to respond to the unmet developmental emotional needs of your inner child and nervous system.

Bonding

A **Redesign** technique that rewires your nervous system for trust and teaches it that your social and emotional needs will be consistently met and validated.

Chapter 7: Awareness

Attuning

to the needs of your developmental nervous system

I have been and still am a seeker, but I have ceased to question stars and books; I have begun to listen to the teaching my blood whispers to me.

Hermann Hesse

Now it's time to take the language of the body even deeper into what makes us human. This is the realm of identity, belonging, and how the security and stability of our sense of self and the world dictates whether your nervous system is regulated or dysregulated. You know that your nervous system is always operating based on its perception of whether it's safe or unsafe, and you now know that it will respond to varying degrees of perceptions of safety with the polyvagal hierarchy of survival responses. But the question remains, why do we feel so unsafe? After all, most of us aren't living in the wild, running away from predators, or desperate for food. By objective measures of physical or mortal danger, most of us are 'safe'. And although the demands of school, work and life are stressful, why do other people seem to be able to navigate the same stressors as us with more ease and resilience? Oftentimes, nervous system dysregulation is the manifestation of wounds to our developmental self or inner child. Russell Kennedy, also known as The Anxiety MD, says 'anxiety is a creative inner child that is trying to get your attention through your body'.

We will explore how early childhood experiences shape our nervous system. In the same way that we explored polyvagal theory to transform your perspective of your body's responses, we will learn how critical early attunement is to our well-being and to shaping our lives. For **Awareness** in Part 3, we will be directing our attention to BASE-H, where H (human) represents unheard messages from your developmental nervous system. You will learn that most of your nervous system dysregulation is in fact a message, in the language of your body, from your developmental self telling you that it needs your *attuning*, *tending* and *bonding*. Once you are able to hear and understand these old wounds and cries for help, you will be able to bring nervous system regulation and healing to the most foundational - and human - of levels, your inner child.

The Developing Nervous System

A baby's brain and nervous system is like a ball of clay waiting to be shaped. That's why there is such an emphasis placed on what parents and society are 'modelling' to our children because their nervous systems literally mimic what we demonstrate. A developing human baby's sensory system is learning from every stimulus, input and experience it has to develop its brain and nervous system. Initially, it gathers information from the mother's womb and her experience. Whatever is going on for the mother, those neurological sensations will be the first inputs into the development of the child's nervous system.[1] After birth, the baby's nervous system is not only recording the inputs from her mother, but from the other people she interacts with as well as all the events and experiences going on around her. All of these stimuli contribute immensely to the foundations of the baby's brain and nervous system and, by extension, who they are and how they interpret the world. To put it very simplistically: If the general experience is one of love, connection, abundance, security and safety, that person's nervous system will by default view the world and her subsequent experiences, even adverse ones, with optimism, security and resilience. On the other hand, if her general experience was one of fear, loneliness, scarcity, vulnerability and danger, that person's nervous system will by default view the world and her subsequent experiences, even the positive ones, with pessimism, doubt and hopelessness. However, for most of us, our upbringings were not as black and white as this. It was a blend of positive and negative experiences. Instances where the experience was that of love, connection, abundance and safety but at other times, perhaps during moments of difficulty and adversity, our emotional needs were either unmet or met with rejection or shame that made us feel unsafe.

Modelling

Long before you are having your own conscious thoughts and opinions, much less actual memories of what happened to you in your life, the most critical programming to your nervous system is taking place through *modelling* and more generally what you are exposed to growing up. Modelling is a psychological concept by which we observe, internalise and emulate behaviours and patterns of those around us, particularly those of our primary caregivers during our formative years. This observational learning extends beyond mimicry. It involves absorbing not only what was explicitly said or done to us but also what remained unspoken, hidden or simply in the background. As young, impressionable individuals, we learn from our environment, making connections and absorbing information at an astonishing rate. Research has even found that higher levels of conflict between parents is associated with infants showing stronger brain activity in certain regions, which has implications for emotional development and mental health.[2] This is seen even when they are sleeping. Our nervous systems, particularly during childhood and adolescence, are wired to pick up on the subtleties of social interactions, and in the process internalise behavioural patterns and coping mechanisms, simply because those are the models we have. Thus, it is not uncommon to unconsciously inherit not only the strengths but also the limitations, insecurities and unresolved issues of our parents, primary caregivers and role models.

The impact of modelling can extend across a lifetime, shaping our relationships, emotional responses and self-perception. You can probably think of some behaviours that may have unconsciously been modelled to you. For example, if a mother frequently exhibited anxious behaviours, such as constant worrying, avoiding certain situations or frequently showed signs of panic, we may have learned to react to stressors in a similarly anxious manner. This can contribute to our own anxiety and difficulty in regulating our nervous system's stress response. Furthermore, we could have learned to respond to frustration or conflict with anger and aggression if we witnessed these emotions frequently being expressed in our home environment. This modelling can lead to difficulties in regulating emotional responses, leading to outbursts and impulsive behaviour. A father who struggles with depression may exhibit behaviours like social withdrawal, loss of interest in activities and a general sense of hopelessness. Growing up in such an environment, we may internalise these behaviours, making us more susceptible to mood disorders and emotional dysregulation. Similarly, if we were exposed to a highly self-critical attitude and unrealistically high standards, we may adopt that outlook for ourselves. This can lead to the stressful ongoing battle with perfectionism and self-criticism to the detriment of our emotional well-being. Parents with substance abuse issues may have modelled unhealthy coping mechanisms, such as using drugs or alcohol to deal with stress. If we were exposed to these behaviours, we may be at greater risk of adopting similar maladaptive coping strategies, potentially leading to addiction and nervous system dysregulation.[3] If we grew up around adults constantly avoiding confronting difficult emotions or problems and relying on denial as a defence mechanism, this may have inadvertently taught us to do the same. And finally, inconsistently responding to stressors, sometimes overreacting and other times underreacting, could have confused our ability to gauge the appropriate stress response leading to difficulties in self-regulation and emotional stability.

Jill *had recovered from a ten-year eating disorder pattern but still struggled with anxiety When the subject of her childhood arose to explore the origin of her coping mechanisms and default survival patterns, Jill said: 'If you think my childhood affected me enough to give me this, I think you're wrong. I don't have any childhood wounds.' We explained to Jill that childhood wounds often develop in silent and hidden ways. One day, in a session, we guided her to observe the impacts of her childhood through BASE, tuning into the experience of the language of her body. Jill had learned to tune into BASE-M but had not had a breakthrough. Today we were tuning into BASE-B. We guided her into the felt sense of her body and a particular sensation was most persistent. She could feel a constant gripping and aching in her solar plexus area. As she remained there a little bit longer, Jill noticed that this sensation would start as anxiety, then travel up to her chest and throat, and make her feel a wave of sadness. After staying here a little longer, Jill began to cry. Once the crying subsided, we asked if there was any thought or voice or tone in the experience of her body. Jill was overwhelmed with confusion and didn't know why she was crying. She said, 'I don't know!'*

We dived back in, inviting her to tune into her body again, and like clockwork that feeling was still there. It was less intense but it was there. After staying here for a couple minutes she quietly said, 'I don't feel worthy ... but I don't know why.' We shared with her that she didn't have to know in her mind, because her body would soon tell her. For the third time, she found that feeling again, which by now was even less intense, but it felt immobile. Stuck in her throat and chest. She whispered, 'You're worthless ...' Something mysterious began unravelling. 'I don't feel anything when I say it to myself ... this isn't my voice. I can hear it and it lives inside me but it didn't come from me. Oh my God! It's my mother's voice to herself. I can see her. I am three or four and she is looking at herself in the mirror crying and saying: "You're so stupid, how could you, you're a failure, you're worthless, you're nothing." I am suddenly being flooded by moments of her being very critical of herself. I know this about her, but I didn't know my body listened or cared.'

Jill became sad. She felt grief for a part of her that needed a secure mother to teach her to be secure within herself. Jill began connecting dots and recognising that although her mother was loving, thoughtful and caring toward her, she was very judgemental and critical toward herself. Clearly her mother was dealing with an unhealed wound too. Jill found immense relief in learning that her anxiety had a root and that the root could repair and heal. Diving into her BASE-H gave her the insights she needed to learn in order to heal the attachment wound that came from her mother's modelling.

Our healing journeys are rarely straightforward and linear but they always reveal layers of hidden treasure, enabling us to mend and nurture the unhealed parts of our mind–body–human and making us more whole. Recognising and addressing unhelpful coping patterns through self-awareness is crucial to breaking the cycle of nervous system dysregulation that can be passed from one generation to the next.

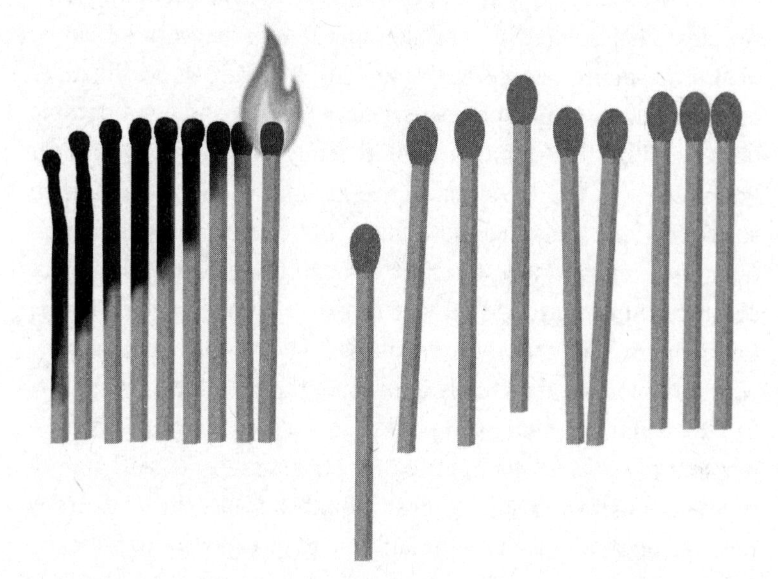

Trauma

As described in the Introduction, trauma can be defined as *too much, too fast, too soon, too slow, too little or too late*. This encapsulates the nature of experiences that can lead to emotional and psychological distress, leaving an imprint on our nervous system. It suggests that trauma is not solely about acute, intense events but also about the absence of necessary support or intervention when it is needed most. This definition acknowledges that one can be deeply affected by overwhelming experiences that occur abruptly or prematurely, leaving one struggling to cope with the lack of attuned, timely or adequate responses to one's needs.

Trauma, therefore, is not just *what* happens to us, but what is not repaired or healed after what happens to us. And while we don't believe that all nervous system dysregulation stems from trauma we do believe that a lot of it does stem from those unhealed wounds. This is being increasingly accepted in the world of public health, as many people begin to see more and more linkages between adverse childhood experiences (ACEs) and adult chronic health issues. The landmark ACE study conducted by Kaiser Permanente in the 1990s set out to assess the level of trauma in childhood and contacted more than 17,000 individuals with questions like: Did you lose a parent through divorce, abandonment, death or other reason? Did you feel that no one in your family loved you or thought you were special? Did you live with anyone who was depressed, mentally ill or attempted suicide?

The results they gathered were truly staggering. More than 60 per cent of respondents reported having at least one ACE and, of those, one in six reported four or more.[4] In short, the ACE study revealed how common it is for children to be exposed to traumatic experiences. Moreover, because the study was conducted by one of the largest health management organisations in the world, the researchers were able to correlate ACE scores with mental and physical health outcomes. For example, the prevalence of depression in people with ACE scores of four or more is 66 per cent compared to just 12 per cent for those with no reported ACEs. Similarly, the likelihood of someone being on painkillers or antidepressants rose by proportionate levels.[5] Adults with ACE exposure have worse mental and physical health, more serious symptoms related to illnesses, and poorer life outcomes. Across numerous studies these effects go beyond behavioural and medical issues, and include higher levels of stress hormones, cancer,[6] reduced immune function and, as we know from our clients, mind–body illnesses like chronic fatigue syndrome, IBS, autoimmune disorders, depression,[7] post-viral syndromes, chronic pain and more.[8]

This research demonstrates how impacts on our nervous system from childhood cause long-term effects, *but* these effects don't have to be permanent. They can be healed and that is exactly what you are doing with this book. But it is important to realise how much your childhood can influence your current state, even if you can't recall a big, traumatic event occurring. Very few people are aware of the impact that early childhood has on them and they tend to either idealise their childhood or marginalise the negative things that happened to them. If you want to heal the wounds of your nervous system, being open to working with the past will help you heal yourself in the present.

Important note: There are many reasons a parent or caregiver might neglect a child and, oftentimes, the neglect is unintentional and unconscious yet can be as traumatising for the child as for the adult. For example, for those of us who may have grown up in poverty with the adult caregiver working multiple jobs, or in a home with a family member suffering a significant, sudden illness over a long period of time, neglecting the children's needs can be unavoidable and does not imply a lack of care but rather a lack of capacity. The wounds for your inner child will feel the same though. Whether the neglect was unintentional or not, it will still impact on your nervous system.

Attachment Is Everything

Throughout this book we have been weaving in the concept of trauma and adversity as likely root causes of your nervous system dysregulation and anxiety, depression and chronic pain - because of how our nervous system develops. And while some of our clients are healing from trauma as it is conventionally viewed - abuse, neglect, abandonment, poverty and violence to name a few - many more are healing from other types of influential experiences that are not taken seriously, not acknowledged or disregarded as 'normal'. We previously discussed how modelling and experiences that we are exposed to as children profoundly impact on our developing nervous system. Now, we will review how the specific quality of our relationships with our primary caregivers (or parents) is critical to our healthy development. In fact, the quality of our relationships, if positive, can protect us from ACEs whereas if the quality is negative, this can amplify our ACEs. Let's dive into Attachment Theory, co-regulation and attunement, but first a story ...

Louis *is a successful and thriving business owner who came to us to help him heal his gut issues and rewire patterns of anxiety and dissociation. When Jen asked him whether he had experienced any adversity in his childhood, Louis replied with a firm no. Louis had the most loving and caring parents, and on the surface his childhood appeared idyllic. His parents showered him with affection and his home was filled with warmth and care. However, beneath this seemingly perfect exterior, a subtle yet significant mis-attunement was at play. One day Louis shared that there was one thing that he thought was worth mentioning. Whenever he sought help with vulnerable emotions like sadness or heartbreak from his father, his role model and favourite person in the world, he was met with* *awkwardness and silence. Rather than receiving the reassurance and connection that he (his nervous system) needed to feel soothed and safe (regulated) in his feelings, Louis encountered silence, which left him feeling confused, inadequate and unworthy. Unbeknownst to him, his nervous system interpreted his father's silence as a form of disapproval, weakness and shame that left him feeling abandoned and defeated. Consequently, Louis's nervous system adapted by frequently oscillating between the freeze and shutdown responses as a way to protect him from his own vulnerable emotions.*

This example illustrates how these types of seemingly normal, mundane and unremarkable mis-attunements in early attachment can have a profound impact on an individual's emotional well-being and nervous system regulation. While Louis's parents provided love and care in the majority of instances, the absence of emotional attunement from his father in times of emotional vulnerability left a pattern in his nervous system leading to symptoms of anxiety and a constant struggle to feel truly worthy and connected. Louis hadn't realised that, although his childhood was wonderful, there was a dynamic that impacted on his nervous system enough to wire it to freeze and shut down rather than regulate. Through attunement, Louis realised that his nervous system adapted to his own vulnerable emotions and his father's withdrawal by deploying a disorganised and avoidant attachment style whereby his survival system was keeping him trapped in freeze and sympathetic mode. By opening himself up to the awareness of those wounds that then felt painful and overwhelming, today he is able to feel attuned to. By listening to the messages of his developmental self (inner child), the 'H' in BASE-H, as his body asking for help, he could then interrupt and redesign his unhelpful patterns from the past. Within a short while, not only was he feeling much more at home in his body and comfortable with his vulnerable emotions, his gut issues cleared up as well.

Many of us may resonate with Louis's experience. Learning that our needs were not met as we needed them to be does not mean our parents don't love us, or that we don't love them. It doesn't necessarily mean we had a difficult childhood or that we are blaming them for it. Learning that our needs were not met as we needed them to be is actually a helpful and useful reflection on how our biology is wired and how it can be rewired - as a first step - through awareness. And as a gentle reflection, our parents were likely doing the best they could with the knowledge they had, just like you up until this point. We usually cope the best we can with what we know, until we learn better. And sometimes even when we learn better, for all the aforementioned survival-nervous-system reasons talked about in this book - doing better is still not simple. It's also important to note that the quality of our attachment bonds in our early years is also influenced by the culture we (and our parents) grow up in, our religious beliefs (and those of our parents), societal norms and long-standing traditions. Sometimes, what was widely accepted isn't what was best for our nervous systems.

Newborns are 100 per cent precious, but they are also 100 per cent incompetent. A baby horse can walk within two hours of being born but a baby human usually doesn't take its first step until it is a year old. The human species has made this functional trade-off for very good reasons. What we are unable to do physically in the first few years, we overwhelmingly make up for with our big brains and unparalleled intelligence later in life. But that trade-off requires human parents to do an enormous amount of work early in a child's life to keep them alive. To ensure the survival of the species and that the care a baby needs is provided, babies are hard-wired to form profoundly close bonds with their caregivers. That's why the one thing that a baby can do immediately upon being born is cry to signal its needs. Put succinctly, the *single most important thing* to a baby's survival is to *form a bond with its parents* so that its needs are met. The absence of parental attention is *literally* experienced by a newborn as stressful and is profoundly dysregulating to their nervous systems.[9] This is why some of the most triggering and painful experiences we can have throughout our lives – as far as our developmental nervous system is concerned – are feelings of abandonment, rejection and loneliness. This means that there is no deeper, no older, no more fundamental root to your nervous system determining if it's safe or unsafe than its perception of how well it's being cared for by its parents. The study of this early childhood bonding and its impact on human development is called Attachment Theory.

It's also critically important to understand that healthy attachment between parent and child is not just based on needs like food and shelter. In the 1960s, Harry Harlow conducted a groundbreaking experiment where he placed newborn monkeys in a cage with two 'mother' monkeys. One was made of wire but provided the baby with milk while the other was made of terry cloth which was soft and comforting but provided no food. Without fail, though the baby monkey would go to the wire mother for food when hungry, the baby spent the rest of the time clinging to the soft and comfortable terry cloth mother.[10] This proved that as far as the baby's nervous system was concerned, comfort and connection (affection and love) were more primary to the baby's needs and development than mere physical nourishment.

A similar phenomenon was recorded almost a century earlier in American and European orphanages. Philanthropists and social welfare advocates created orphanages to house and care for abandoned babies but many of the newborn babies died within one year. It was eventually determined that the reason for these tragic statistics was the fact that the facilities were understaffed and the caregivers only had enough time to feed and clean the babies. The babies were dying from a lack of attention, interaction and touch. As staffing levels increased and the babies were cared for beyond their most basic needs, the death rates dropped.[11]

As soil and rain are to a plant, food and water are to a baby – but the true source of a plant's health and growth is the light of the sun. Without it, the plant will wither and die. Similarly, a parent's attention, affection and responsiveness are like the sun for a baby. All of Attachment Theory boils down to the fact that close, emotionally attuned attention, affection and responsiveness from a parent to a child is developmentally essential for the healthy and optimal function of human beings.[12] If the baby's perception is that it can always count on its parents for protection when it's scared, for touch when it's in pain and for love when it's lonely, its nervous system will generally perceive itself and the world as safe and secure and therefore be regulated. But if the opposite is true, if when it's scared it is not comforted, if when it's in pain there is no comforting touch and when it's lonely love doesn't come, that baby's nervous system will generally perceive itself and the world as unsafe and insecure and be perpetually dysregulated.

Co-regulation and Attunement

In its most basic meaning, attunement is our ability to be aware of and respond to a child's needs, particularly their emotional needs, relational needs (feeling seen, accepted, included) and needs for safety. Co-regulation is what happens when attunement is met by helping the child regulate their emotions and reactions, particularly in stressful or challenging situations. Over time, co-regulation of the nervous system teaches a child self-regulation of the nervous system. This is important because, like a newborn child being unable to walk for some time after birth, babies are also born without the ability to regulate themselves, so they literally depend on others in their life to help them to do so.

The best way to understand this is to examine it in action. Take a moment to imagine a four-year-old running around on the playground who trips and scrapes their knee, feels pain and fear, and begins to cry and search around for help. Now imagine how that child would feel after the following interactions:

1. Their mother approaches, stops a foot away from her child and inspects the wound and says: 'Let me see your cut. Oh that bruise isn't that bad. You're OK. It's nothing to worry about. Stop crying. It's not a big deal.' And then she waits with her kid until they quiet down and tentatively start playing again.
2. Their mother approaches and immediately embraces her child and says: 'I'm here, sweetie. That looks like it really hurts. I'm sorry, baby. It's OK, I'm here with you now.' And then she holds her child until they stop crying, take a big breath and then wriggle out of her arms to begin playing again.

Scenario 1. describes mis-attunement and a lack of co-regulation. The parent focuses on the physical injury instead of the child herself or her emotional needs. Moreover, instead of empathising with her child's experience, she condemns her feelings of pain and fear by telling them that what they are feeling isn't necessary and that they should instead stop crying. The lesson the child learns from this is that her very own feelings are wrong and shameful and that to make Mom happy she needs to suppress them and pretend to be OK. The child's nervous system goes from sympathetic activation to a functional form of dorsal shutdown to repress her feelings and appease her mother.

Scenario 2. describes attunement and co-regulation. The parent focuses on the child herself. She embraces her and acknowledges her emotional state of fear and pain and lets her know that feeling those things is OK *and* that she'll be OK. She stays with her child and helps her co-regulate by being a source of warmth, supportive contact and soothing tones until the girl moves from sympathetic activation and naturally completes the cycle back to ventral vagal. In fact, oftentimes once the emotional and survival-driven responses are met with emotional and somatic co-regulation by the parent, the child returns to calm and without any attention paid to the physical boo boo (because it was never really the problem) she's ready to play again.

If scenario 1. is repeated over years throughout the young girl's early life, she will have learned to suppress her own feelings in an effort to cope with the pain and hurt that comes from the lack of attunement from her mother. If you've had experiences like her, you will likely have done the same. Moreover, as a result of being mis-attuned to as a child, you become mis-attuned to yourself. Not only are you unable to provide for your own primary emotional and relational needs, you *don't even know you have them!* And that is what this chapter is all about, expanding your awareness skills to include *attuning* so that you can decipher the secret language of your body and hear the long-forgotten messages, needs and shouts for help from your development nervous system (inner child).

Attachment Injuries and Coping Patterns

Any child that suffers from chronic mis-attunement will have what are called attachment injuries. Attachment injuries by necessity create coping patterns in the nervous system to try to conceal and work around the injury. In the same way that you know someone has an injury if they walk with a limp, when we see someone exhibiting compulsive people-pleasing behaviour or intense perfectionism, we can reasonably assume they have an attachment wound from childhood. These coping patterns present along a spectrum of behaviours that attempt to secure attachment like people pleasing or, on the other end, attempts to 'not need' secure attachment through hyper-independence or avoidance of close relationships. We say 'not need' because secure attachment is always an absolute necessity of the nervous system. That being said, the nervous system can suppress its own needs to such a degree that a person can authentically believe they don't need other people. This may be functionally and practically true, but the absence of human connection still has significant impacts on the quality of their health and life. These patterns of behaviour in babies were classified by attachment researchers in the following styles: secure, anxious, avoidant and disorganised. The styles manifest in adult life as follows:[13]

Secure: Able to form lasting and stable relationships. When relationships are under stress or rupture, someone with a secure style will feel some dysregulation but not be overwhelmed by it. Furthermore, they will be attuned to their own needs as well as those of others.

Anxious: Able to form relationships but is continually anxious about the security of the relationship. They need frequent reassurance and will usually compromise their own needs and desires for others out of fear of being abandoned. When relationships are under stress or rupture, their dysregulation presents as high levels of anxiety, desperation for reassurance and people pleasing.

Avoidant: Has difficulty forming relationships, being aware of their own feelings and sharing them with others. Exhibits self-sufficiency and independence. When relationships are under stress or rupture, their dysregulation presents as distancing and isolating themselves.

Disorganized: The rarest style and usually the result of a traumatic childhood. They have difficulty forming long-term and stable relationships. They will exhibit a combination of anxious and avoidant behaviours and often act very intensely, erratically, unpredictably and sometimes dangerously. They may have other mental health disorders related to their trauma.

Though knowing your attachment style can be useful, your style is not an indelible aspect of who you are, it is simply a set of behaviours that have been given a label and, as you are learning in this book, those behaviours are symptoms (just like anxiety, chronic pain or IBS) containing vital messages from your nervous system about what it needs. In other words, your 'attachment style' is just one more presentation of your nervous system deploying strategies to try to survive a threat. In this instance, the threat is compromised attachment that, as noted above, is perceived by your nervous system as life or death. As with the polyvagal hierarchy of responses, most people present with a blend of attachment styles and coping patterns. We are going to discuss some common presentations in the stories that follow and explore how their patterns, nervous system dysregulation and attachment wounds fit together. As you will see, by learning the secret language of the body and directing their awareness to attune to their fundamentally HUMAN attachment needs (rather than the behaviours covering up those needs) clients found their way to the deepest level of healing.

How Attachment Injuries Manifest Later in Life

By all measures Karden is a high-performing person, always has his act together, isn't needy and is known for being there for others. For most of his adult life, Karden assumed he was independent, capable and obsessed with taking care of people because that was his nature. The only real chink in his armour was low back pain. And over the course of his mid-twenties to mid-thirties, despite being a trained bodyworker, functional movement therapist and having access to the best of the best in care, his pain became more frequent and more agonising. It was when he was on the brink of having surgery that he stumbled upon John Sarno's book *Healing Back Pain*, which professed the idea that chronic pain is the result of personality styles and repressed emotions (or as we say, nervous system responses). Sceptical but desperate, Karden took his first dive into listening to the secret language of his body.

As Karden probed the pain in his back using the listening skills from Chapter 1, he discovered that there was more to feel than just pain; there were the sensations of clenching, of both holding on for dear life but also holding something in at all costs. And underneath that, much to his surprise, were massive waves of rage and sadness. As he allowed the waves of emotion to be felt, he experienced an instantaneous reduction in his physical pain. It was as if the pain was being caused by emotional pressure on the dam of his back, and once that emotional pressure was felt and released, the pain was no longer needed. Then Karden took his enquiries into his body even deeper.

Using the framework of attunement and inner child work, he imagined that his low back was his child self and that all of the feelings were those of his inner child. He then asked his inner child why it was feeling all that rage and sadness. And what came back were two essential messages. The first message was with the rage and said, 'I hate having to take care of everyone all the time.' The second message was with the sadness and said, 'Why doesn't anyone ever take care of me?' In that moment, Karden realised for the first time that there were repressed parts of himself, stemming from early childhood, that desperately wanted to be taken care of. Although at first glance he had a fine childhood, when he looked below the surface and past the default idealisations that most people place on their parents, things had not been as good as he 'remembered'. He grew up in a divorced home and though his mother had been wonderful in many ways she was profoundly neglectful in others. She carried immense intergenerational trauma and attachment injuries of her own, was extremely self-centred, obsessed with work and was severely addicted to prescription opioids (that began from treating back pain), which contributed to her premature death when she was only 54 years old.

Although she had provided him with food, shelter and Christmas presents, and was quite affectionate when she was present, on the whole she was not present and the lack of time and attention was devastating. Karden was left alone for huge swathes of time, he was frequently left waiting for a ride home from school, he had to feed himself most of the time. His mother rarely made time for his interests, barely showed up to important events and would without fail place her needs ahead of his. Her neglect had caused an attachment wound in Karden where his child self had learned that he didn't matter, that his needs didn't matter, and that hurt terribly.

But no child can function with that much pain and so Karden's nervous system repressed those feelings and invented ways to avoid dealing with them, to receive attention and not depend on his mother. Those became his coping strategies of having his act together (he would repress his own needs and vulnerabilities so he didn't have to feel the horrible feeling of not mattering), independence (if his mother couldn't provide for him, he would provide for himself) and taking care of others (if he couldn't get the secure attachment he needed, he would take care of others and get snippets of attention in the form of approval for doing good things, which was better than nothing). And ultimately, like many others, his coping strategies had become his identity.

By listening to the language of his body, Karden was able to delve the depths of his humanness and reclaim his connection to his child self and form an embodied attuned relationship with him. Using the skills of tending and bonding, the goal was not to extinguish the behaviours of independence, competence and taking care of others. After all, those are useful traits. The goal was to heal the attachment wound so that those behaviours were no longer driven by emotional injuries from childhood and corresponding dysregulation. Karden's back pain became a welcome and reliable (though unpleasant) message from his nervous system for when he was neglecting himself, and would cue him to connect with his body and inner child.

Charlotte *was an actress, model and photographer who had been working tirelessly at her career for more than a decade. She was highly motivated, worked crazy hours and was a self-proclaimed perfectionist in everything she did – and she was perpetually anxious. In 2021 during the Covid-19 pandemic she left the city where she grew up and worked, and moved to Florida. The pandemic along with the move were overwhelmingly stressful for Charlotte, and she promptly crashed and became housebound with severe POTS, IBS, pain and anxiety. A few months later she joined our programme.*

Using listening and translating skills, Charlotte connected with the messages that were underlying her symptoms. Without fail, behind every symptom was an incredible amount of raw fear. She had always known she was anxious, but she was dumbfounded to feel the intensity of the fear that was simultaneously immobilising and shaking every part of her. At first, she couldn't make sense of it. Like most clients, when the first time I asked her about her childhood she said: 'It was fine. I wasn't starving and my mom loved me.' But as we worked together and used attune techniques to connect with her inner child and re-examine her past, it all began to make sense. When Charlotte asked her scared inner child why she was afraid she said back to her: 'No one is here to take care of us. I'm afraid we're going to die.'

As it turns out, Charlotte and her mother were abandoned by her father when she was very young and they had housing and financial trouble. Her mother (because of her own attachment injuries) was unable to cope, showed all of her fear and shared all of her worries continuously with Charlotte. Charlotte learned from her mother that they were unprotected, that the world was a dangerous place, that her mother was terrified and that all they had was each other.

No child can endure perpetual fear so, like Karden's, Charlotte's nervous system repressed those feelings and invented ways to cope. In this case, she decided to be 'the perfect girl' in all things so she did not worry her mother and she became an overachiever. Her nervous system reasoned that if she was perfect at everything, presumably nothing bad would happen to them and everyone would like her. She also established a reverse attachment role with her mother where she felt she needed to care for and comfort her instead of the other way around.

Despite her coping patterns working for many years, her inner child and deepest layers of her nervous system lived in a perpetual state of danger. When Covid happened, all her worst fears were proven true. The world was an unsafe place, there was no escape, she was socially isolated and couldn't be with her mother; her coping skills of perfectionism and overachieving could do nothing to protect her (after all, she couldn't even work) and her nervous system collapsed.

Charlotte's healing began when she stopped looking at her POTS, small intestinal bacterial overgrowth (SIBO) and anxiety as problems with her body, but as signs from her nervous system that her inner child was in a state of terror and dysregulation. Whenever her symptoms were present or she noticed herself going into her old patterns of anxious perfectionism, she would listen to the fear and attune to the unmet needs of her inner child. By bringing connection, support and love as well as helping her inner child change her view of the world, Charlotte redesigned her brain's habits of thought and feeling, healed her attachment wounds and eliminated her symptoms. She has returned to her career, works hard and still brings a perfectionist's eye to her projects, but she's no longer driven by fear.

Your Inner Child

An old African proverb says that *the child who is not embraced by the village will burn it down to feel its warmth*. We hope that you can now see from the lessons and stories in this chapter that ACEs and attachment wounds are at the heart of how your nervous system moves you through life. Our inner child is burning down our nervous system in order to get our attention. In order to heal these traumas and wounds from your past and regulate your nervous system in the present, you must work on *attuning* yourself to the child-like and developmental language housed in your body that are behind your triggers. As with the previous two sections of the book, following A I R, we will first support you in becoming aware of what is going on with the attuning practices that follow.

Remember, we can't effectively heal what's dysregulated inside us if we don't know what's actually happening. That's why we start with awareness. Because so much of healthy attachment is the co-regulation that occurs simply from a parent being present and caring for their child, when you do the attuning techniques, your compassionate and caring Observing Self will be providing co-regulation to your inner child simply by listening to what it's communicating through BASE-H and its verbal messages. With the connection formed through *attuning* and the sacred messages you receive from your inner child, you'll be ready to interrupt your coping mechanisms with the *tending* techniques (Chapter 8) and redesign a new relationship with your inner child using *bonding* techniques (Chapter 9).

> **Your nervous system's survival patterns are the body's response to your inner child's wounds.**

Let's take a moment to summarise the knowledge on Attuning to the needs of your inner child for Awareness:

- The most critical programming to your nervous system takes place through *modelling* as you grow up.
- Our nervous systems, particularly during childhood and adolescence, are wired to pick up on the subtleties of social interactions and internalise behavioural patterns and coping mechanisms.
- Witnessing inconsistent responses to stressors, sometimes overreacting and other times underreacting, may confuse our ability to determine appropriate stress responses, leading to difficulties in self-regulation and emotional stability.
- Childhood impacts on our nervous system can cause long-term effects *but* they can also be healed.
- Attunement is the ability to be aware of and respond to a child's emotional needs; even the needs of your own inner child can be met and attachment injuries can be healed.

Practices 7
Awareness: Attuning

Learn to attune to the developmental needs of your nervous system

In these practices you will identify your attachment symptoms, correlate them to your survival responses and discover your underlying developmental needs by attuning to your inner child.

Why Do It?

The attuning techniques will help you learn the real reasons for your nervous system dysregulation.

When to Do It

When you are triggered or dysregulated and want to become aware of the attachment-based pattern that needs attention.

Tips before you begin

- Curiosity and being an objective observer of your experience will help you see the truth and logic underlying your unhelpful behaviours.
- Avoid self-judgement, criticism, rationalisation and justification regarding what you observe.
- Your inner child's needs are simple, raw and powerful, and are verbalised at a reading level of a 9-year-old or below. The simpler and more direct the messages are the more accurate. Avoid complexity and abstraction.

Practice 1: Identifying Your Attachment Symptoms (Style)

Reference the chart on page 261 to assist with this practice.

1. Find a quiet place to sit in.
2. Take a deep breath.

3. Tune into the body with BASE.
4. Think of a recent argument or relationship rupture that you have experienced.
5. As you think about that, notice what happens in your mind and body.
6. Use your Observing Self to notice your thoughts, feelings and actions that you exhibit during such experiences.
7. Map what you observe onto one of the attachment symptom patterns below.

Repeat this practice a few times on relationship triggers (like arguments or hurt feelings). Continue to compare your description to the guide below to attune yourself to your default attachment symptom pattern. Note them in the developmental coping pattern rubric below.

PS: You may ask your relationships (family, friends, lovers) how they think you tend to act during ruptures to gain more insight.

Practice 2: Attuning to Your Developmental Needs with BASE-H

Remember that your attachment style is actually a presentation of coping behaviors (symptoms) that are trying to resolve an underlying developmental wound. Now that you have identified your style, healing comes from identifying the developmental message and addressing the wound.

In this exercise we will be guiding you into identifying the developmental wound of your inner child which is the HUMAN of BASE-H. What you will find is that the root of disregulation you

	Secure	Anxious	Avoidant	Disorganized
Internal Working Model	I know that I matter and am worthy of love. This hurt is temporary and will be repaired soon	I'm not worthy of love. I can't trust anyone to love me or care about my needs	When my partner is upset they must think that I am a defective failure	Nobody can be trusted. No one is safe. Everyone is attacking me
Wounded Feelings	My feelings are hurt but I know that I am safe	I feel so rejected, sad and abandoned. They hate me and everything is awful	I feel rejected and abandoned but I am so dissociated I can't even feel it	I feel rejected, abandoned and terrified. I am in danger
Protective Feelings	My upset is real but it's safe to feel this and it will pass	I am angry! If they cared about me they wouldn't do this to me. Or I'm afraid. I need to make them happy so they're not mad at me	I am hurt and angry but I can't feel it, so I am going to act as if I am OK even though I am not	Angry, anxious, volatile, stuck
Behaviour	Take some time to reflect, talk about it, repair and reconnect	Attack, blame, demand my needs be seen and validated. Or people please, appease, fawn	Ignore it, pretend it's OK, passive aggression, silent treatment, isolate myself	Act erratically, fight, flee, go silent, manipulate

Attachment symptom presentations

are experiencing in BASE is almost always coupled with -H (the developmental wound of your inner child).

1. Find a comfortable place to sit.

2. Use your imagination to visualise yourself in a place (like your childhood bedroom, classroom or playground) with your Observing Self and a child version of yourself, approximately 3–8 years old.
3. Bring to mind a recent argument or relationship rupture you have experienced.
4. Observe how your inner child reacts.
    a. What's their facial expression?
    b. Identify their posture and body language.
5. Notice and label the actions, sensations and emotions that are present.
6. Knowing that your inner child's needs are always raw, simple and powerful and that they communicate at a very basic level, ask them to share what they are feeling with you. You may use the chart below to help you determine what primary need(s) they are expressing to you.

Now that you're aware of the developmental messages coming into your mind, and body, listen and label them using BASE-H.

1. **Breath:** What am I feeling in my breath?
   E.g. I am holding my breath, my chest feels contracted and painful.
2. **Action:** What does my body want to do and how does it want to move?
   E.g. My body freezes and doesn't know what to do.
3. **Sensation:** What am I feeling in my body and where?
   E.g. I am feeling a pit in my belly and a feeling of heaviness and numbness in my arms and face.
4. **Emotion:** How am I feeling and where?
   E.g. I am feeling fear in my whole body, especially my chest.
5. **Human:** What is the raw, simple and powerful message from your developmental self (inner child)?
   E.g. I feel hurt, lonely and sad. I want to curl up in a ball.

Message	Developmental Need	How to Provide It
Feels hurt, lonely and disconnected	Attention	Need to be seen and spent time with
Feels scared, anxious, state of survival	Protection	Need to feel safe, protected, embraced
Feels hurt, lonely, disconnected, scared, anxious	Belonging	Needs to be included, surrounded by attention
Feels empty, insecure, worthless and depressed	Validation	Needs to feel like they matter by being genuinely seen and told that they are special
Feels painful, lonely, abandoned, worthless, defective	Warmth	Needs comforting, soothing and affectionate physical contact and engagement

Unfiltered inner child messages (reference chart)

Make Attuning a Part of Your Life

The goal is to quickly get to the root of your dysregulation by accurately identifying the childhood threat instigating your survival response. To do this, we invite you to:

1. Observe and notice how most of your triggers have a developmental origin.

2. Use your Observing Self to check in with your inner child whenever you find yourself triggered.
3. Become more and more familiar with your patterns so you can tend to them faster.

Notes:

....

....

....

....

....

....

Chapter 8: Interruption

Tending

to our human

What happens when people open their hearts?
They get better.
Haruki Murakami

With the awareness skills of *attuning* you can now decipher the messages from your inner child through the language of your body. This awareness makes it possible for you to use *tending*, the third interruption technique in A I R, to respond to those messages and bring regulation to the deepest levels of your nervous system. As discussed in the last chapter, these messages are calling for co-regulation through attention, protection, belonging, validation, warmth and responsiveness. Until the nervous system receives those things in an authentic, embodied and consistent way, it will continue deploying old coping mechanisms from childhood. Fortunately, the wonders of neuroplasticity make the healing of developmental injuries and trauma possible. By using the tending techniques, you will be able to finally change the conversation in your mind and body about your perception of self and your relationships, and learn to feel safe, loved, worthy and connected. By extension, this will regulate your nervous system and substantially advance your healing from anxiety, stress and unresolved trauma.

In order for the process of tending to work, it must connect to the developmental aspects of your brain, what we call your inner child. Pioneering clinicians like Margaret Paul, the creator of Inner Bonding Therapy, and Richard Schwartz, the creator of Internal Family Systems, have created therapeutic approaches that allow people to access the developmental, relational and social realms of the nervous system. Both approaches recruit imagination, visualisation and dialogue to allow your Observing Self to connect with your inner child. In A I R, we combine elements of both these practices and enhance them with a deep connection to the language of the body and polyvagal exercises to provide you with a simple, direct and effective method for healing the aspects of your physiology that we call your Human. In the paragraphs that follow we will review some essential concepts about the inner child, the Observing Self and the language of your body, as well as introduce another dimension of the polyvagal theory so that you have everything you need to begin tending to your younger, unhealed self.

Inner Child Needs

The needs of your inner child are the universal needs that every developing human has. Determining what the needs of your inner child are isn't a guessing game or a mystery. As we talked about in Chapter 7, important needs are: attention, security, belonging, validation, connection and responsiveness. A child needs to feel like they exist, feel safe, feel included, feel valued, feel loved and to have their communication (non-verbal and verbal) acknowledged and reciprocated. This means that any coping pattern or dysregulation that you are experiencing in the present, if related to childhood (and it usually is), is *always* related to one or more of

Need	Description	Injury	Coping Pattern	Polyvagal
Attention	Need to be seen	May feel unwanted, unloved, lonely	May act out or behave in exaggerated and unexpected ways to get attention	Fight
Security	Need to feel secure and safe	May feel unsafe, abandoned, neglected, unable to self-regulate	May experience control and rigidity, hypervigilance, excessive mood swings, withdrawal	Flight, fight, freeze
Belonging and Acceptance	Need to feel accepted, included in the family or group	May feel left out, hurt, lonely, disconnected, anxious	May experience isolation, people pleasing, difficulty trusting, takes things personally and social anxiety	Flight, shutdown, freeze
Validation	Needs to be seen, heard, requires self to be positively reinforced by others	May feel empty, insecure, unworthy, depressed	May experience overachieving, perfectionism, acting out, exaggerated emotion, withdrawal	Fight, flight, shutdown
Connection	Need to feel connected, comforted, soothed, support, companionship, meaningful communication	May feel lonely, abandoned, unworthy, 'not enough', defective	May experience people pleasing, perfectionism, overachieving or coldness, isolation and avoidance	Fight, shutdown, freeze
Response	Need to feel needs being met, connection and provision of preceding needs	May feel uncared for, hurt, worthless, anxious, doubtful, insecure	May experience clinginess, excessive need for reassurance or withdrawal, isolation and avoidance	Flight, shutdown, freeze

these categories. Each individual is different and part of your work is to notice which of the categories tends to drive your dysregulation and trigger you the most.

When it comes to attuning to the needs of their inner child and noticing triggers, people tend to abstract, overthink and complicate what their inner child needs. The simpler you can make it, the better. Four-year-olds don't rationalise or abstract their lives, they feel them in the unfiltered binary of safe or scared, included or rejected, having or not having, loved or abandoned. When we are adults, the situations that trigger our inner child may not immediately appear as clear-cut as the binary of safe or unsafe, but on closer inspection they always wind their way back to an attachment wound related to one (or more) of the development needs. On page 273 is a framework designed to help you see each of the developmental needs, what it may feel like in your body when they are lacking, and some of the potential coping mechanisms that result and their polyvagal survival category. The table is by no means all-inclusive of the possible coping patterns that can be displayed. In order to begin tending to yourself, you need to use our *attuning* skills to observe your patterns and connect the dots between the coping mechanism that signals that your inner child has been triggered and the specific need that's not being met.

Listed above are some of the different variations of one single giant need - love. Love nurtures the social bonds and connections that sustain us. It provides us with resilience and wholeness through a solid foundation and profound sense of attention, security, belonging and acceptance, validation, connection and responsiveness. Love offers a safe haven where feelings, thoughts and behaviours can be freely expressed, integrated and redirected when necessary. So love isn't just love, and it's once again not as black and white as loved or unloved. It comes in all the shades and colours described above. Distinguishing which categories you resonate and identify with can be helpful in learning more about the human inside you that is begging for their needs to be met - that can help you heal. As you become more familiar with your patterns, you will be able to address the individual developmental needs calling for your attention with clarity and compassion, and be able to regulate yourself swiftly and effectively. However, nothing worth doing happens overnight. Since this is a process of getting to know oneself, particularly aspects of self that have been hidden and repressed since childhood, it takes time and we invite you to be kind to yourself as you get to know your inner child.

Note on unmet needs and consequences on health: You can see how, in the secret language of the body, unhealed trauma and unmet needs lead to coping patterns that create a stress state in our overall physiology. When our nervous system is repeating survival states of fight or flight, shutdown and freeze, our cells, organs and body systems are also in survival mode and not thriving at their best. Chronic states of stress (with a nervous system stuck in chronic fight, flight and freeze mode) has been found to lead to chronic illnesses stemming from different systems and parts of our body like cardiovascular conditions,[1] gastrointestinal disorders, psychological disorders, metabolic disorders,[2] reproductive health issues, respiratory conditions[3] and autoimmune diseases.[4]

Self Co-regulation

In Chapter 3 you learned how to access your Observing Self and began using it to witness your own experience and initiate change within your nervous system. Practising Observing Self was also your first foray into exploring the helpful idea that your identity and conception of self is not a singular 'I' but a consciousness composed of many selves. As it turns out, our mind and body are quite adept at occupying multiple selves or 'parts' as they are called in Internal Family Systems Therapy. In fact, we are doing it all the time. An example of this is when you say things like, 'Part of me wants to quit my job, but another part of me wants to stay.' Presumably you are one person but in this instance, like many instances in our lives, we have multiple viewpoints and conflicting desires. Our experience of ourselves is therefore indeed not a singular 'I' but a collection of parts that represent the diversity of our thoughts, opinions, feelings, coping patterns, survival mechanisms and nervous system responses to our life. When we adopt this more helpful conception of ourselves as a family of parts, it makes it possible for us to use our Observing Self to work with ourselves much more effectively. It can connect with the competing forces inside our mind and body, facilitate dialogue and ultimately foster a family of parts within ourselves that are consciously working together rather than unconsciously competing for survival and control.

In this chapter, your Observing Self will take on one of its most important roles in your healing. Your Observing Self will become an attuned adult and respond to the needs of compassion and care, and begin offering attention, security, belonging, validation, connection and responsiveness to your developmental self (inner child). Tending involves using your imagination to visualise a relationship between your attuned Observing Self and your developmental self to provide the co-regulation that your nervous system needs to feel safe. We call this self co-regulation. For this co-regulation to be truly effective it needs to include the language of the body in addition to words and images.

The Wordless Relationship

Albert Mehrabian, Professor Emeritus of Psychology at the University of California, Los Angeles, is known for his research on the communication of emotions and how we know if we like or dislike a person. He determined that when people receive emotional communication, the meaning is conveyed not only through the words being said, but the tone and body language of the speaker. He also determined that only 7 per cent of the meaning conveyed in emotional communication is through the words themselves, whereas 38 per cent is conveyed through tone and 55 per cent is conveyed through body language.[5] That means that 93 per cent of emotional communication between humans - whether adults, adolescents or children - is non-verbal!

Imagine you were tasked with comforting an upset baby, but you weren't allowed to make eye contact with her, touch her, hold her, rock her or even make soothing sounds to her. All you were allowed to do was to use words. If your immediate thought was 'that would be impossible', you are absolutely right and it means you understand that attachment, connection and co-regulation between a baby and its caregivers are mediated through the somatosensory system, especially touch, and that words in this case are in fact useless.

Contact is an essential human need. When Jen's securely attached one-year-old son Leo is introduced to a new environment, especially with new people, his nervous system protects him like most children by keeping his distance from the new people and clinging to Jen for safety. His little nervous system depends on his close physical contact with her to feel secure. With the security of that embrace, Leo is then able to observe his mother's positive emotional connection with the strangers and as his nervous system registers that she trusts them by monitoring her movements, facial expressions and tone of voice, he learns through co-regulation that the new people are friends. Within an hour, even though he likely doesn't understand all the words being spoken, he will go from being cautious and distant to feeling safe and secure. This need for non-verbal connection continues to be true even after children have mastery of language.

Karden's five-year-old daughter Leia has both excellent language skills and a precocious intellect, but in moments of distress, especially when there is a rupture between the two of them, words, explanations and rationalisations are rarely effective. If something she has done has frustrated him, she can see it on his face and she'll immediately get upset and say, 'Don't look at me that way!' And as long as Karden is still feeling frustrated and that emotion is being broadcast by his face and body, even when he says, 'It's OK. I'm sorry,' his daughter doesn't feel soothed. This is because her senses are finely attuned to the language of the body and she knows that the words aren't true. In order for Leia to feel safe, she needs to know that her connection with her father is secure. It's only after Karden takes a few moments to feel beneath his frustration, find connection and shift his own nervous system back into love that his body language and facial expressions change. Then he can look at Leia, she can see and feel that he loves her and her nervous system co-regulates with his.

You've learned throughout this book that your body speaks a wordless language rooted in breath, actions, sensations and emotions, so it may not surprise you that the same wordless language is how one human truly speaks to another, how a mother really connects with her child. Therefore, healing our own inner child requires us to lean into that same wordless meaning to direct co-regulation at ourselves. A beautiful example of this in action came in a letter from one of our students named Jordan.

Dear Jen and Karden
I had a huge breakthrough with little Jordan last week! It happened (of course) after a major activation and breakdown, but as you guys say, every breakdown is an opportunity for a breakthrough. Anyway, last Thursday my wife came home from grocery shopping. As she was unloading the bags I asked her if she remembered to get the ketchup that I had asked her to get. She said, 'Oops, sorry, I forgot,' and I instantly got angry and said, 'I specifically asked you to get ketchup!' She said, 'I know, I'm sorry, I forgot, I had a lot on my mind, but it's OK, we still have ketchup, we haven't run out.' And I said, 'That's not the point, I wouldn't forget to get you something that you had specifically asked for.' Then she got angry too and said, 'First of all, it was a mistake. Second, WE STILL HAVE KETCHUP, so it's not a big deal. And if you forgot something I asked for, I wouldn't flip out on you.'

It was at that point that a part of my Observing Self kicked in and though I was still really angry, my OS was saying, 'She has a point. We still have ketchup. It's really not a big deal ... Why are we so angry?' It took a huge amount of effort, but I told her that she was right and that I needed a few minutes to work through what I was feeling. So I went to our bedroom, laid down and took a few minutes to notice what I was feeling. There was so much anger and constriction it was hard to think. It took 10 minutes of downregulating NSMs to get to a point where I felt regulated enough to summon my inner child. I could see him in my mind's eye, curled into an angry ball in the corner of the room (just like my own son does when he is angry). I could see my inner child and feel his anger in my body. At the same time I was able to look at him as my attuned observing self and feel the parts of my body that felt calm, curious and compassionate towards him and his anger. I asked him if I could move closer to him and he said, 'No.' So I asked him if he wanted to be left alone, he hesitated and then said, 'No, please stay.' For a minute or two all I did was stay in the vision with him, a few feet away, focusing on the parts of my body that held feelings of calmness and sending it to him. As I did that, I could feel the parts of my body holding anger soften and I saw the contracted ball of his body softening, opening up and slowly turning towards me.

I then asked him if it would be OK for me to ask him a question and he said, 'Yes.' So I asked him, 'Can you share with me what's making you so angry?' and he said, 'If she loved me, she wouldn't have forgotten the ketchup.' My OS found it astonishing that my inner child and nervous system was reacting with such drama and intensity to this simple act of forgetfulness, but I also knew that there had to be more to it and I started to draw some throughlines to my childhood. So I asked him, 'Have there been other times that you've felt this way?' All of a sudden I was teleported to being 7 or 8 years old, sitting on the floor in front of my couch, curled up in a ball and crying. I was so incredibly sad. It was just one of hundreds of times that my mom hadn't come home to be with me and make dinner when she said she would. I had called her at 5:30 and she said she'd be home at 6. She wasn't. I had called her at 6 and she said she'd be home by 6:30. She wasn't. Called her at 6:30 and she said she would be home by 7. She wasn't. I was hungry, I was alone, all I wanted was my mom and all I could think and feel was that ***if she loved me she'd be home to be with me,*** *but she wasn't.*

I let the waves of sadness that were underneath my anger move through my body. I cried and I sobbed as I laid in my own bed as a

35-year-old man. At the same time, I had my hands on my heart and my belly and I could feel that they were the hands of my attuned observing self. At some point I asked my inner child if he wanted me to hold him and he said, 'Yes.' I let him crawl into my lap in the vision and I could feel our bodies connecting. His body began to conform to mine and I could feel warmth and connection in my body as well as the sadness. I told him that he had every right to feel the way he was feeling and that it was wrong of our mom to do that to us. I told him that I loved him and I poured my love and warmth into him while he was in my arms as well as through my own hands on my heart and belly into myself. After a little while all I could feel was calmness and presence between the two of us. I told him that I would always be there for him. That we are no longer a little boy. And we had a little chat about how our wife was not our mom and that she loves us and cares for us all the time.

As he and I stayed in that calm place and I helped him see the truth of what our life was like in the present, it was as if he was seeing it for the first time. As if he had been stuck in some kind of time capsule and could only see the world through how he felt in the past. But after this impromptu session of inner child work, he now could see things as they really are. That he is loved. That we are loved. Since then, whenever my nervous system has been activated by similar triggers, I've been able to quickly feel my body, connect with little me and remind him with both my words and my body that I am with him, that I love him and that we are loved by lots of people. I settled very quickly. Honestly, it feels like I have superpowers ...

Jordan's letter demonstrates fluency in the secret language of the body. He has fully attuned to his mind-body-human presence using aspects of the A I R approach, including:

- Overall *Awareness* skills to notice his dysregulation
- *Interruption* skills to take some time to disrupt his nervous systems default responses
- *Listening* skills to notice what was actually happening in his body underneath his activation
- *Modifying* NSM techniques to regulate himself
- *Attuning* techniques to discover the needs of his inner child
- Visually rich and embodied *Tending* techniques to provide for the needs of his inner child to bring profound regulation
- And, to be discussed more in the next chapter, he applied *Bonding* techniques to redesign his relationship with his inner child

Social Engagement System

As you are learning in the third section of this book, our nervous system's healthy development and sense of safety is utterly dependent on its sense of social connection and inclusion (attachment) to other human beings. As a result, human beings have evolved aspects of our brain and nervous system that exquisitely monitor social and emotional cues from our fellow humans. In fact, another one of Stephen Porges' contributions to science was identifying the role the vagus nerve plays in transmitting and regulating social cues of safety and danger in the body. He coined the concept of the Social Engagement System (SES) as a component of the polyvagal theory.

Let's recall the ventral vagal state from Chapter 4. In that state we feel calm and comfortable, curious, connected to others, mindful and present. Our body feels relaxed, restored and able to comfortably engage in any social, physical or psychological activity. Your nervous system's capacity to be in the ventral vagal state is governed by its determination that it is socially and relationally safe and it does this through assessing the facial expressions, tone of voice and body language of those around you. If a person walks into your office and you see a warm smile, bright eyes, relaxed muscles, open posture and hear happiness in their voice as they say hello, your brain's SES will interpret all of those signals as positive. Simultaneously, the vagus nerve will send messages to your own body that things are safe and comfortable and it will instruct your own face to smile and your eyes to widen to transmit back to the person similar signals for friendliness and receptivity.

Conversely, we are all familiar with getting 'bad vibes' when we meet someone and find ourselves saying, 'I can't put my finger on it, but I have a bad feeling about that person.' Even though your cognitive brain doesn't know why, your embodied SES does. Perhaps your brain detected that the posture of the person was rigid, his walk had a touch of aggression in it, his eyes, even though looking at you, weren't really seeing you, or maybe the shape of his mouth indicated that he was in a miserable mood. Upon tracking those cues, your vagus nerve transmitted that information to your own body indicating that there might be a threat of some kind. In doing so it may have tightened the muscles around your neck, inhibited your breathing and generated a small pit in your stomach. In fact, it's only because of the sensations happening in your body that your mind knows that something is off. As usual, the body leads and the mind follows.

Referring back to the example of Karden and his daughter. When he became angry with her, her body only relaxed when her SES detected that Karden's mood had actually shifted through his facial expressions. The SES is so fine-tuned and so influential that a simple change in a facial muscle or shift in tone of voice can either regulate or dysregulate us. Fortunately, we can put this to our advantage when we incorporate the SES and the regulating power of the vagus nerve into our tending practice.

You may have heard of the research finding that making yourself smile (even when you aren't happy) can elevate your mood.[6] The reason why this is possible is because the vagus nerve is a two-way highway that transmits information from the brain to the body and vice versa. When you smile your SES makes a positive association and the vagus nerve tells the brain, 'We're happy.' And then the brain sends reinforcing messages back into the vagus nerve which starts to impact the rest of your body, making you feel more energised and deepening your breath. It's this two-way street where positive inputs initiated in the body can stimulate our SES and move us toward a ventral vagal state that can greatly enhance our tending techniques.

Being The Parent You Always Needed

The human section of this book and the tending techniques you are about to learn are really about cultivating the attuned Observing Self inside you to become the parent your inner child always needed. To do this, you will be using all the skills you've accumulated in this book and the attuning techniques from the last chapter to find, connect with and bring attachment and love to your inner child through the language of your body. By repairing your earliest wounds you will facilitate some of the most profound healing possible and open up whole new possibilities in your life.

The child that you were when your trauma and attachment injuries occurred did not have the development, maturity, knowledge or life experience to successfully navigate them. But the adult you are today, with her accumulated life experience, wisdom and strength, that is committed to healing and armed with the knowledge of A I R, can navigate those wounds and rescue your inner child from the past she is stuck in.

Let's summarise the knowledge we've covered on Tending to your inner child for Interruption:

- Tending to your inner child will help change the conversation in your entire being, and help you feel worthy, connected, loved and whole.
- To help you change the conversation it's necessary to learn what the conversation is about through tuning into the unmet needs of your inner child.
- Your Observing Self will help you become an attuned adult, fluent in the secret language of the body by learning how to self co-regulate.
- The language in your human is wordless, and thus requires a fine-tuned approach.
- This approach can be explained through the lens of the social engagement system.
- Tending will help you become the adult you need yourself to be today by healing the adult you needed when you were younger.

Practices 8

Interruption: Tending

Learn to tend to your developmental needs and speak the language of attachment

In these practices you will learn how to provide attuned attachment to your inner child which will regulate your nervous system from the ground up.

Why Do It?

Tending techniques regulate the root of our dysregulation by bringing safety and resolution to our oldest traumas and attachment wounds which we have been trying to manage with coping mechanisms our entire lives. When we regulate the foundation, we regulate everything above it as well.

When to Do It

Anytime you are triggered and the trigger is related to a developmental wound such as abandonment, inconsistent and unpredictable care or conditional love (which is highly likely) and you feel disconnected to yourself and the people in your life.

Tips before you begin

- As with the attune techniques, the needs of your developmental nervous system (inner child) are raw and simple – always. Try not to over-complicate it.
- The language of the body and BASE are far more effective means of tending to your inner child than words.
- All you need is love.

Practice 1: Polyvagal Social Engagement Exercises

Note that the following exercises are to be practised with a partner or loving friend.

Engage in Connection

1. Find a person you feel safe around.
2. Ask them whether they would be willing to share a few moments with you in this powerful exercise that will benefit you both.
3. Once consent is obtained, find a comfortable place for you both to sit together.
4. Take a moment to notice how you feel, and invite your partner to do the same.
5. Sit side by side, with your right arm and their left arm touching.

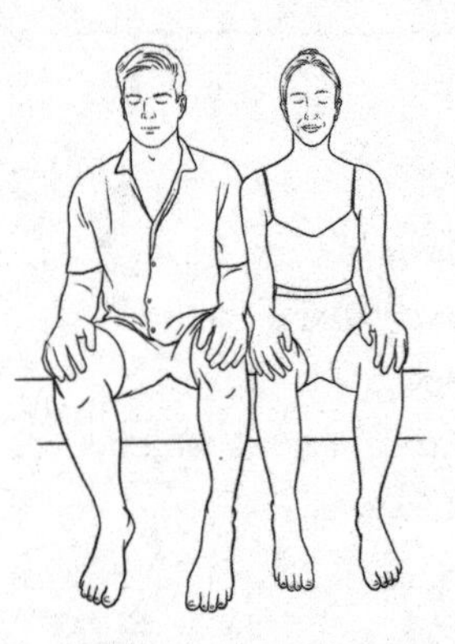

6. Stay here for as long as feels comfortable and notice what it feels like to feel the caring warmth of this person.
7. Move your arms away from each other and allow yourself to smile, sigh or move in any way that feels comfortable. Invite your partner to do the same.

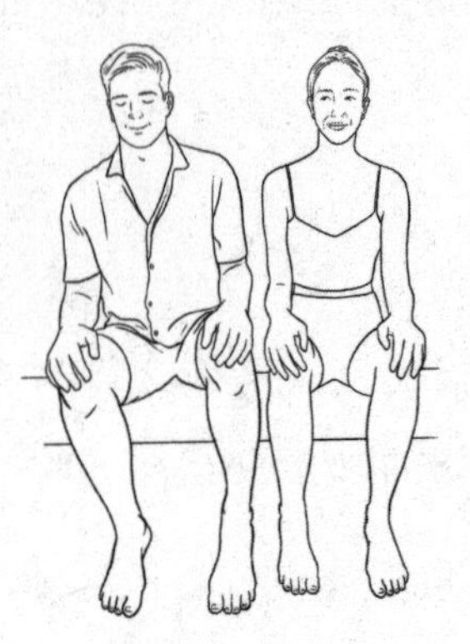

8. Repeat the exercise with your left arm and their right arm touching and notice the difference.
9. Check in with yourself and observe how things have already shifted inside you.
10. Invite your exercise partner to do the same.

Engage in Co-soothing

1. Find a person you feel safe around and connected to. Maybe even do this exercise with the same person you practised *connection and warmth* with above.
2. Ask them whether they would be willing to share a few moments with you in this powerful exercise that will benefit you both.
3. Once consent is obtained, find a comfortable place for you both to sit in together.
4. Take a moment to notice how you feel, and invite your exercise partner to do the same.
5. Sit back to back in a cross-legged position with your backs touching.

6. Gently begin swaying from side to side feeling the contact with the other person's back.
7. Notice how it feels to be supported in this way. Prompt your exercise partner to notice as well.
8. Stay here swaying for as long as feels comfortable.

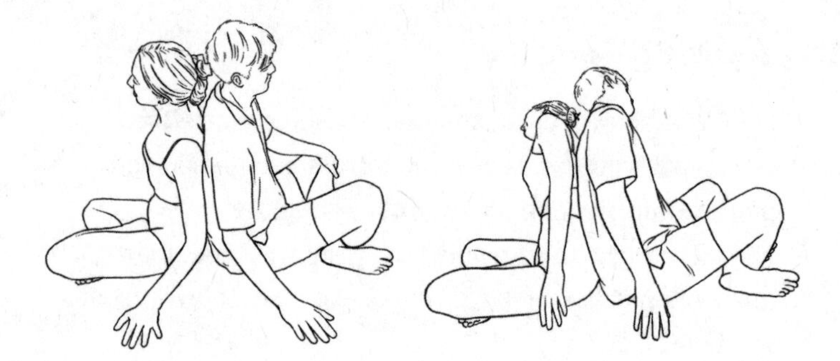

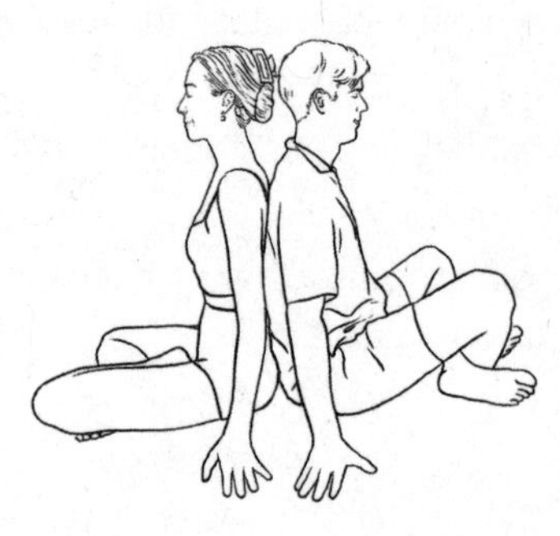

9. Move your bodies away from each other and allow yourself to smile, sigh or move in any way that feels comfortable. Invite your exercise partner to do the same.
10. Check in with yourself and observe how things have shifted inside you. What does it feel like to feel the soothing experience of supporting each other with gentle movement and rhythm?

Practice 2: Providing for Your Inner Child

Part 1: Attune

1. Find a comfortable place to sit.
2. Use your imagination to visualise yourself in a place (like your childhood bedroom, classroom or playground) with

your Observing Self and a child version of yourself (approximately 3–8 years old).

3. In your visualisation, observe the state of your inner child.
    a. What's their facial expression?
    b. What is their posture and body language?
4. Notice and label the actions, sensations and emotions that are present in their body.

Part 2: Connect

1. Now imagine yourself there, with your inner child, as your Observing Self and the qualities of the attuned adult: presence, attention, care, compassion, protection and love.
2. Feeling these qualities in your body, look over at your inner child and imagine that you could broadcast these feelings and qualities to them.
3. Notice perhaps your facial expression softening, your head tilting and your eyes smiling. Your inner child's experience is likely already shifting. Notice this shift.
4. Ask your inner child to share with you what they are feeling and what they need.

Part 3: Provide

1. In your embodied visualisation, provide what they asked for through the language of the body. For example, if they said they feel like no one cares about them, imagine your inner child sitting in your lap and feel the contact of their body against yours, wrap yourself in a containment hug (Chapter 6) and imagine gently wrapping them in an embrace and feel it, speaking out loud, and with a tone of voice that

carries authenticity, have your Observing Self let them know you care about them deeply.

2. Notice and label how this feels.

3. Intuitively provide attuned attachment to your inner child within this embodied visualisation for as long as feels comfortable.

Sometimes your inner child will only be able to tell you what they're feeling and what's wrong, but not what they need. In that case, it's your job to be intuitive and experiment with providing what they need. If it feels accurate (or inaccurate) the language of your body will always let you know.

PS: It can be helpful to imagine ideal parent figures from real life or fiction (like Pixar movies or books) to inspire you in becoming the best parent you can be to your inner child.

Make Tending a Part of Your Life

The goal is to make tending to your developmental needs a primary practice for regulating yourself. To do this, we invite you to:

1. Observe and notice how most of your triggers have a developmental origin.
2. Attune and tend as often as you can.
3. Take your time in getting to know your inner child and forming a loving relationship with them.

Notes:

....

....

....

....

....

....

Chapter 9: Redesign

Bonding

with your nervous system

I'll love you forever,
I'll like you for always,
As long as I'm living,
my baby you'll be.

Robert N. Munsch, *Love You Forever*

In Part 1 of this book, the Redesign practices taught you to cultivate your Observing Self as a new and more helpful mode of mind. In Part 2, the Redesign practices taught your nervous system to make settled states more familiar, longer lasting and, over time, your new normal. Now, in Part 3, we apply the third step of A I R by demonstrating how to make *bonding* with your inner child your new default. We don't mean to suggest that any of us wilfully ignored our inner child, but before learning the secret language of the body, even just noticing the pain and hurt was impossible. When our nervous system became dysregulated by a developmental trigger, the only thing it could do was deploy any of its numerous coping mechanisms to help us get away from feeling our own hurt. As long as we continue to ignore the pain of our inner child, our body will continue to scream out messages, and that pain will run our lives. On the other hand, when we commit ourselves to leaning into that pain and to provide attuned, attached and compassionate care to our inner child, our pain becomes a powerful portal to enriching and expanding our life.

This chapter is all about making the default become your attuned adult speaking the secret language of the body through consistent and responsive attention to your developmental nervous system. Since so many of your nervous system's survival strategies revolve around your unmet developmental needs, training yourself to be consistently engaged with them will take your healing to a whole new level. We will explore the concept of our 'internal working model' and how it governs our perspective of ourself, our, relationships and the world. We will also review the power of neuroplasticity and see how by regularly *bonding* you can transform your entire nervous system's perspective from one of surviving to thriving. Finally, we will introduce the concept of anchoring, which will help your brain and body more rapidly shift into bonding. But before we dive in, we wanted to share with you a few examples of the power of this work.

After 12 weeks of participating in our courses, we invite our students to write a letter to the version of themselves that started the journey three months earlier. To illustrate some of the transformative healing that can happen with bonding, below are a few of the letters students have shared with us:

Dear Francesca
Deep down inside you always knew you would heal, but you didn't know how. Thank you for choosing yourself and taking on this work. Despite your mother, father and friends not understanding, you persevered and listened to your body, a language that was foreign, a voice that was so quiet and ignored you didn't know it was there. The voice that believed it was too much and could not show itself to the world. And yet, you courageously stood up for little Franny. Would you believe me if I told you your symptoms are the unheard

voice of our inner child guiding us to a life you are meant to live? If I told you that you don't need anyone else but yourself to heal? No more waiting for someone to prescribe you that magic herb or drug that will make your exhaustion go away. For someone to take you to the doctor, only to have your hopes dashed and labelled with chronic illness with no cure. For someone to finally see you as a whole and complete Francesca, not a sad victim. You are so much more than a label and an illness.

You are no longer a prisoner in your own body. The cell door is open, and what you have learned is that it is OK to step out and to trust our attuned adult when she says that you are not guilty or bad. We now know that getting sick was not our fault, but it is our responsibility to lead our healing. Every day, you are learning it is safer and to take up space in your body, to not just survive but to thrive, and be thankful for everything: including symptoms because they all contain sacred messages. I know that sometimes the fear feels all-consuming and the weight of life too heavy. I know when it feels like that, you want to give up, but what we're also learning is that endings come with new beginnings. You're beginning a new life, one where you have all that you need inside you, where you are not the victim, where you are healing yourself, taking leadership of your life and choosing a new path of listening to your body, healing your inner child and creating a new life. You are a self-healer who chooses to do the things that are hard with grace, because you have everything you need inside you and nothing can stop us now.

Dear Maddy
I don't know where to start. So much has transformed. It's easy to think of all the work I still have to do and practise, but holy shit, reflecting on the past couple of months is amazing! I have so much more awareness now than I did before. Looking back I can see how perfectionism and people-pleasing dominated my life and motivated

everything I did. Sitting with my inner child and listening to her, holding her, being with her, compassionately observing my triggers coming from my childhood. Making the unconscious conscious. You can't unlearn this stuff. There's no going back. And thank goodness.

I'm prioritising my health while also meeting my inner child with patience and compassion. It's a fine line to walk. My love for little Maddy grows more and more every day. I can feel how badass I am now I feel like I know myself now more than ever. That I've tapped my intuition in a whole new way. Choose what works for me and what doesn't. Trusting that my observing self and attuned adult can heal me. Trusting that I know what's best for me and not needing to outsource or seek external opinions or validation 24/7. Allowing myself to make mistakes. Allowing myself to slow down. It's easy to look at the mountain of healing that's in front of me, but then I remember to look back at everything up until this point. To take one step at a time. To focus on how far I have come rather than how far I have to go. After consistently using my skills, I am healing. I am not broken. Proof to keep going. I know I am on the right path.

Internal Working Models

The term internal working model (IWM) was popularised by John Bowlby, the pioneering researcher and creator of Attachment Theory. Every child develops an IWM, which is a mental representation that governs their perception of themselves, others and the world. In addition, their IWM will become their inner guidance system for making predictions about how interactions with people and life events are going to go. In the same way that we develop an IWM that touching a hot frying pan will hurt us after being burned, a child will develop a much more complex and life-shaping IWM about whether they can trust people based on whether they are treated in an attuned or mis-attuned way by their primary caregivers. If a child's experience is that when they are in distress their parents don't give them enough attention and attuned care, or worse, their needs are shamed or ignored, they may develop an IWM similar to the hot frying pan, except in this case they learn that *people* can burn them. Because a person's life is dependent on interactions with their fellow human beings, the impact of someone's IWM on their life cannot be overstated.

In addition to how they view others, a child's early life experiences determine their IWM of themselves. Human beings are meaning-making machines. Our minds are constantly striving to not only make sense of our body's experience, like we addressed in Part 1, but make sense of the world around us. To know why things are the way they are. When a child is raised with secure attachment where their primary caregiver provides ample and reliable amounts of attention, security, belonging, validation, connection and responsiveness, their minds automatically make sense of that experience. The mind of a child who receives a lot of love and has their needs met says to itself, 'I must be a good and worthy person to receive all this kindness and care.' Therefore their IWM of themselves (identity) will reflect that they are worthy, that they matter, that their needs matter and will be confident speaking up and advocating for themselves.

When a child does not have their needs met when they are in distress and their needs are abandoned, or shamed, their mind must make sense of why they are being treated that way and they say to themself, 'This must be happening to me because I am no good, something is wrong with me and people don't like me.' Consequently, their IWM of themselves (identity) will reflect that they are unworthy, that they don't matter, their needs don't matter and they will have great difficulty speaking up or advocating for themself because when they do, no one cares. When our IWM is organised around unworthiness, mistrust, fear of others and overall we view the world as a dangerous place, our nervous system is incapable of truly feeling safe and employs coping patterns that keep it stuck and chronically dysregulated. But through tending to our inner child through our nervous system, we can repair our IWM and extraordinary healing is possible. *Bonding* is really about being an attuned adult and re-shaping your IWM from one that is governed by worthlessness and fear to one of profound worthiness and love.

A Tale of Two Internal Working Models

Karden had a client in his mid-thirties named *Zach* and they had been working with triggers around male authority figures. Between sessions, Zach had an important recollection and realisation. He came into Karden's office and shared that his parents had separated when he was nine months old. At the same time, his mother began a relationship with Paul, who would become his stepdad. Zach has a rich and loving connection with Paul as an adult, but as a child that was not always the case. His stepdad was, at first, a stranger. He was over six feet tall, enormous and intimidating. In addition, Paul was overworked, constantly stressed, was frequently in a bad mood and had a temper. Zach's most vivid memories of his stepdad were moments when his temper boiled over. In one instance, he remembers being in the living room and something he had done had enraged Paul. From his seven-year-old perspective, it was as if Paul had transformed into a monster, his face distorted, eyes bulged and his voice thundered. Zach was utterly terrified and, as was usually the case, his mother was not there to protect him. Frozen with fear as his stepdad stalked over to him, Zach's survival instinct kicked in and he ran away. But his stepdad was bigger and faster than him so as Zach rounded the corner and climbed the stairs to escape, his stepdad got a solid strike to his butt before letting him get away. Zach then locked the door to his room, hid under his covers and cried till he fell asleep. As a result of experiencing many instances like this throughout his childhood, Zach's IWM was that Paul was safe to be around when he was in a good mood, but if he were in a bad mood, he was invincibly strong and terrifying. Zach avoided conflict with Paul at all costs because his embodied nervous system knew that if Paul got angry, horrible things would happen. As a result, his nervous system went on to apply that IWM to other male authority figures in his life. Even as an adult in his thirties, his body responded to potential conflict with older males with fear and shutdown. But then Zach recalled a remarkable experience when he was 21.

As it happened, Paul's temper had cooled in Zach's late teens and by the time he went to college he hadn't interacted with Paul's rage in many years. Then one day, Zach was with Paul, his sister Penelope and his stepsister Sarah (Paul's biological daughter) and a heated family argument began. At a certain point, the rage monster that Zach had not seen for many years erupted from Paul with volcanic force at Penelope. Zach remembers his terror, his body freezing, the blood draining from his face and even remembers his 21-year-old eyes orienting to the staircase where he so often had made his escape from Paul as a child. But then he saw something that he couldn't believe. In the face of Paul's invincible rage and strength, Sarah stepped in to defend Penelope. She stood her ground and roared right back at Paul and told him that he couldn't yell at her like that. Zach was stunned. What he was seeing, as far as his IWM was concerned, was impossible - that someone could be in the presence of Paul's wrath and not be destroyed.

He told Karden that at the time he couldn't make sense of it. Worse yet, not understanding the nervous system dynamics at play, Zach had criticised and shamed himself for feeling so cowardly (as a fully grown man) in the face of Paul's temper while his stepsister had been so brave. But now he realised that it was because he and his stepsister had two completely different internal working models of their relationship with Paul. Whereas he had never felt safe because Paul was initially a stranger and occasionally hit him, Sarah, on the other hand, by virtue of being his biological child as well as a female, had experienced a gentler Paul who had never struck her in anger. Therefore, Sarah's IWM was that Paul loved her and that even when he was angry, she was safe.

With this realisation, Karden and Zach summoned his attuned adult and invited his inner child and 21-year-old self into a visual-somatic dialogue. The first beautiful thing Zach's attuned adult did was to help his 21-year-old self release his shame and perception of himself as a coward. With the self-criticism out of the way, Zach was then able to listen to his inner child's fear and sadness. He tended to him by embracing him, letting him feel how big and strong they were now as an adult, reminded him of their martial arts training and assured him that they could protect themselves. By making this a daily practice, Zach was able to begin to reshape his own IWM of conflict with male authority figures. Whenever Zach could feel his body deploying his childhood fear and survival mechanism, he would remember to use anchoring in real time to rapidly connect with his inner child and remind him of their maturity, strength and ability to protect themselves.

Beyond Habit Change: Neuroplasticity and Perception of Self and the World

There is no fixed IWM. There is no 'this is just the way I am'. The brain and nervous system are the most changeable parts of your biology and they are capable of infinite possibilities. In *Buddha's Brain*, Rick Hanson and Richard Mendius cite the following remarkable facts about your brain:

- Your brain is three pounds of tofu-like tissue containing 1.2 trillion cells, including 100 billion neurons. On average, each neuron receives about 5,000 connections.
- The number of possible combinations of 100 billion neurons firing or not is approximately 10 to the millionth power, or 1 followed by *a million zeros*, in principle; *this is the number of possible states of your brain* [emphasis added]. To put this quantity in perspective, the number of atoms in the universe is estimated to be 'only' about 10 to the eightieth power (1 followed by eighty zeros).
- Conscious mental events are based on *temporary coalitions of synapses* [emphasis added] that form and disperse – usually within seconds – like eddies in a stream.[1]

At the risk of oversimplification, the IWM we developed through childhood is simply our most commonly used and repeated *temporary coalition of synapses* that represents our perception of ourselves, others and the world. It is the well-worn path in the woods that we described in earlier chapters. Because neurons that fire together wire together, those temporary coalitions of synapses become very well-worn paths in our nervous system and feel like they are fixed. Yet there are *10 to the millionth power* possible states your brain could be in. And as Zach's story above demonstrates, we can create new states and new working models, a truth that thousands of our other students can attest. David Eagleman, one of the foremost neuroscientists in the world, says that our brains are not hardwired, they are *livewired* and are literally reweaving themselves all the time. Our job is to transform this from a random process into an intentional process.

As Karden likes to say:

> A lifetime of countless random acts wired your neurons into their present formation which constitutes your default self. Therefore, a carefully selected set of intentional acts can rewire your neurons into a new formation that gives you a new self of your own design.

The way you create your new self, your new IWM and new neural pathways is by using the same principles that created your current wiring: repetition, intensity and time.

- **Repetition:** The brain will learn whatever it does over and over again. Within the context of healing attachment injuries, repetition goes hand in glove with the attachment principle of responsiveness. The more interactive a parent is with their child in attending to their needs, the more that child's brain learns to expect to be cared for and feel secure. Therefore, when practising bonding, repetition matters not only because it's the bedrock of neuroplastic change, but because it mimics the critical concept of responsiveness from development.
- **Intensity:** Intensity refers to the degree of authenticity, embodiment and vividness that we experience in our bonding practice in BASE-H. The more you can feel how

attention, security, belonging, validation and connection feel in your body as you bring them to your inner child, the more you'll enforce new neural pathways. Being in a playful, fun state is one of the more powerful states for practice.

- **Time:** The longer the period of time the brain does something or is exposed to something, the more it will become wired. The developmental process that gave you your current IWM began in utero and extended throughout your childhood. Therefore, the longer you practise bonding with your inner child, the more those changes will become ingrained and become your new default.

Whereas in the earlier chapters we discussed these elements in the context of changing the habits of your mental and survival patterns, we are now using them to forge a new relationship with our inner child rooted in those key needs for development to reshape your IWM and regulate your nervous system. Since it was developed as a child attempting to survive and explain adverse childhood experiences, the way we transform the IWM is by consistently interrupting the old patterns when triggered, providing the attuned response and using bonding to reinforce (over and over again) the new IWM that our inner child is safe, worthy and loved. This reinforcement must occur both in the mind through a new narrative about yourself (we are loved, we matter, we are worthy) as well as being deeply felt in the language of the body (BASE-H) so that the autonomic nervous system transforms as well. Over time your inner child will shift from expecting abandonment to expecting support.

> **There is no permanent version of yourself.**

Bonding

The heart of bonding is the consistent provision of attuned attachment over time. In the following pages, we will share some concepts that help amplify the effectiveness and speed of healing your inner child, transforming your IWM and regulating your nervous system.

Now Is Not Then

You will recall that your nervous system is a pattern recognition and habit execution machine. By its very design it deploys the same set of responses developed in the past to respond to any similar stimuli in the present. In the story of Zach above, his nervous system fitted any 'older male authority figure' into the pattern of his stepdad and then deployed the outdated and unhelpful survival responses that he used as a child in his present adult life. Essentially, anytime we get triggered, we get pulled into our past. If we are not aware of this and do not have the skills to counteract it, all of our experience, maturity, intelligence, wisdom and adultness goes out the window and our mind and body are being operated by a nervous system who thinks we are still *that* child.

Recognising the universality of this phenomenon in clinical practice, a common form of support that trauma therapists provide to their clients is to help them realise that *now is not then*. This first happens on the mental level, using the Observing Self to notice and remind the rest of us that 'now is not then, that there is nothing wrong with what we are feeling right now, but it's the response of our child self from the past'. Next, we can start to somatically counteract the time warp in BASE-H by using our somatosensory system to orient ourselves to our present surroundings. This is as simple as using elements of the five senses meditation to look around the space you're in and feel your body in the here and now.

Finally, with your Observing Self distinguishing the past from the present and your sensory system grounding you into the now, you can imagine inviting your inner child to be with you in the present, the very space you are in, with the adult version of you. In doing so, you are making a choice to pull your inner child into the present and co-regulating, rather than allowing them to drag you into the past. By bringing the trigger and response into the present, you maintain access to your adult self and all its experience, maturity, intelligence, wisdom and nervous system regulation skills. Your child self in the past does not have the resources to navigate whatever is triggering you, but your adult self who can speak the language of the body sure does.

> **When you get triggered, your nervous system gets sucked back into the past. In order for this not to happen, it has to learn that now is not then.**

Jen had a client named *Elsa* who suffered from anxiety and chronic fatigue syndrome. She experienced bouts of rage and loneliness, followed by bouts of sadness and shutdown. Together they worked with A I R through Elsa's mind and body. Elsa was particularly responsive to NSMs because she could shift her state and improve her symptoms quite significantly – a very motivating experience to have when healing from a chronic illness. Jen knew that Elsa's developmental wounds were the real driver of her dysregulation but she also wanted to give Elsa's nervous system time to stabilise with the mind and body aspects of A I R first. Over a few weeks, Elsa's window of tolerance widened considerably and her symptoms subsided as well. At this point, Jen felt confident to support Elsa in going deeper and guided her through BASE-H. They identified that she was holding her breath, and that she felt stuck. They found a pulsating sensation in her arms, tension in her jaw and more pulsating achiness in her head. Elsa was able to find anger and fear behind these sensations. As they tried to quest deeper, Elsa's Observing Self tried to connect with her developmental needs and discovered that a part of her was afraid to connect with her inner child, almost as if meeting her would mean feeling the worst hurt of her life. Jen gently invited Elsa to use some settling NSMs and with those her freeze response softened and she found a way in. Elsa began sharing that it was as if her inner child had been curled up in a ball all this time, trying to soothe herself. Jen invited Elsa to find the basic unmet need (that unhealed wound). All of a sudden Elsa said out loud, 'Why don't you love me?' and began crying. Jen invited her to use the butterfly hug, the containment hug and the vooo sound. This gave Elsa enough of a window to become curious about that question that came up for her. Elsa began recalling moments where her mother would dismiss her when she needed help and shout at her when she cried. She felt lonely and completely dysregulated. Elsa said: 'I want to give little Elsa a hug. I want to tell her that I love her. That she deserves me to love her.' Jen invited her to visualise and feel herself doing just that. A week later Elsa was a completely different person – she was attuned and connected with herself. She reported having been triggered and using anchoring to rewire her inner child's response and teach her nervous system that then isn't now. Now we are a safe, capable and skilled adult. Jen invited Elsa to write a letter to her inner child to further deepen the bond between her and her developmental nervous system.

> Dear little Elsa
> At first I ignored you, and I'm so sorry. As we began this journey to heal your wounds, I could never have imagined the depth of hurt you were in. Each step I've taken, each moment I've spent in reflection, has been a building block in reconstructing my sense of self, repairing the hurt within you and me. We are not unlovable. We are not unwanted. And now, as I write this, I can see the once-fragmented mosaic come together to reveal a beautiful and complete picture. The whole me. The most impactful shift has been in the way I relate to you. I've learned to listen to your voice, hold you close and nurture you with the love and compassion you so deserve. I'm sorry that's not what you got when you needed it most. Through this process, together, we've uncovered layers of hurt, so many layers. Thank you for being patient with me as I learned to address them with tenderness and care. Thank you for being patient with me as you helped me connect the dots between your past and my present, revealing patterns you always knew were there, but that I could not yet see. Identifying these patterns and where they lived in your body - and mine - changed the way I perceive myself and the world around me and gave me everything I needed to feel in order to heal. In these last months, I've learned to prioritise our relationship, our bond. I've met you with understanding and love, acknowledging the worthiness you've always possessed and always deserved. Thank you for showing me how strong I can be but also how vulnerable I can be. Vulnerability is what taught me how to tend to the chasm in your heart that all these years caused me to be anxious, tired, angry and sick. Now I know that chasm didn't need filling, distracting or avoiding. It needed healing so that it could repair itself from the

inside. Every day, I witness myself stepping more fully into my authentic self, thanks to the connection I have with you.

With love and appreciation, Elsa.

Elsa recovered from her mind-body symptoms and became a mind-body coach. She is thriving in her new embodied self.

Resourcing Attachment

Your nervous system speaks BASE and BASE responds to memories and visualisations that are rich in images and sensory information. In the same way that recalling a bad memory triggers a cascade of physiological responses like constriction in the chest, a pit in the stomach, or clenching of a jaw, a positive memory can do the opposite: a deep breath, a softening of the shoulders, an expansion of the heart. Any memory or sensory-rich visualisation that generates a positive response in the language of the body is called a 'resource'. You'll remember that in the *Listening* contrast exercise in Chapter 1, the positive experience you called to mind was, in fact, a resource. Zach used embodied sensations of his own strength and skill as a martial artist as a resource to access and shift him into a more regulated state when he was triggered by authority figures.

When it comes to resources for attachment, it's enormously helpful to gather a collection of positive relationships that contradict the negative aspects of your IWM in regards to people and how they treat you. These positive relationships can be with a friend, spouse, teacher, coach, therapist, co-worker or even the barista at your favourite coffee shop who always remembers your name and your order - it can even be a pet. Each resource you can connect with is an opportunity to help your inner child modify its IWM and what to expect from others. Moreover, resourcing your connections with others, even pets, simulates co-regulation, stimulates the ventral vagal response and is deeply regulating.

Write out a list of the positive relationships in your life both from the past and present. They do not have to be 'perfect' or deeply personal, they just have to be people in your life that treat you well and show you elements of attention, security, belonging and validation, connection and responsiveness. And then identify a particular way or instance in which they demonstrated that they cared about you.

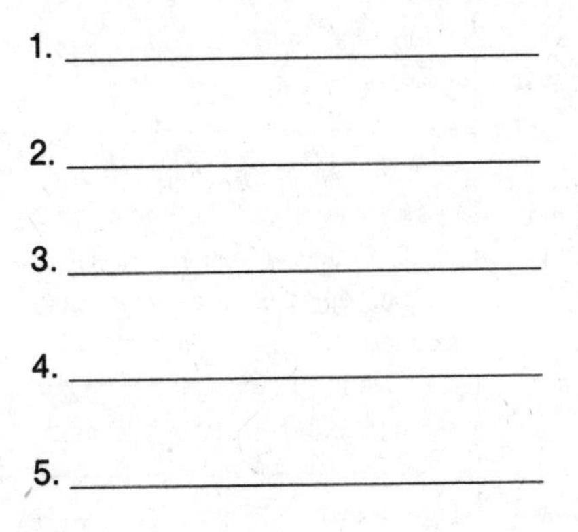

Updating Your Story

Adverse childhood experiences and attachment injuries create a story in the mind and the body about yourself, others and the world. The nature of being a human being is that we live our life and react to others and the world not as it really is but from the prison of our story. For example, someone can have a loving spouse, children and lots of wonderful friends, but because they grew up with a severe deficit of love and attention from their parent figure, they still feel deeply anxious, insecure and like no one cares about them. The reason for this is that our IWM, which doesn't know the difference between the past and the present, will usually warp current reality to conform to its story about the world from childhood. Consequently, it becomes our responsibility (and opportunity) to use your Observing Self and be an attuned adult to show our inner child the truth of life in the present.

You may recall in the last chapter when Jordan connected to his wounded inner child he wrote that as he and his inner child stayed

What's happening	**Story led by inner child**	**Story led by the Observing Self**
The basic facts of the situation	*The reactive, default interpretation of your inner child*	*The new and more useful interpretation led by your Observing Self*
My wife said, 'You were late for family dinner again, can you try to be on time?'	My wife never appreciates me and she has no idea how hard I work.	She does love me. In fact, the reason she wants me home for family dinner is to spend quality time together. Let me remember and feel all the ways she shows me love for a moment. It's OK for her to express her needs to me.
I sent an email to my boss 24 hours ago and he hasn't responded.	I must have done something wrong, he doesn't like me and I'm in trouble.	Let me take a deep breath. I know he has 1000 emails in his inbox and is very busy (just like me). I can recall how pleased he was with the results of my last project. Let me remember all the ways he appreciates me. And let me set a reminder to email him again if I don't hear back in another day.
I wasn't invited to a friend's party.	She must not care about me.	It's more likely that she simply forgot. It's OK. She invited me out to lunch just last week. She does care about me – let me remember all the ways she shows this.

in that calm place and he helped him see the truth of what life was like in the present, it was as if Jordan was seeing it for the first time. As if he had been stuck in some kind of time capsule and could only see the world through how he felt in the past. This is exactly what it's like when you begin to separate the story your nervous system is always projecting onto your life and instead choose to interrupt it and redesign it with a more accurate, healing and regulating story grounded in the reality of your life. And this is exactly what it's like when you speak the language of your nervous system. In any given moment, a very useful tool for doing this is to use the following framework to sort out the facts of what's happening, your inner child's story vs what your Observing Self's story could be about it.

We invite you to mark this page, or take a picture of this table. It may be helpful for you to fill this out when you notice your nervous system pulling you into the story of your inner child to help you recognise what is *actually* happening.

What's happening	**Story led by inner child**	**Story led by the Observing Self**
The basic facts of the situation	*The reactive, default interpretation of your inner child*	*The new and more useful interpretation led by your Observing Self*

As an added bonus for updating your story you can try what we call 'diamond hunting'. It can be really helpful to make a practice of perceiving all the goodness in your life, big and small. The goodness can come from anywhere; for starters, it can come from the resources you listed above regarding relationships. But it can also include all sorts of random goodness and serendipitous events, like having found this book, how you met your best friend, got your job or found a skilled therapist to work with. This is similar to a gratitude practice, but it is less about being thankful and more about counteracting negative narratives that no longer serve you by *finding goodness in your life and feeling that goodness*. To do it make a list of all the diamonds you can think of. Then use your Observing Self to deliberately place your attention on the positive experiences in your life and taking the time for those things to be felt in BASE. Doing this over and over again whenever you find yourself being sucked into your old narrative is a critical part of transforming your IWM.

List your diamonds

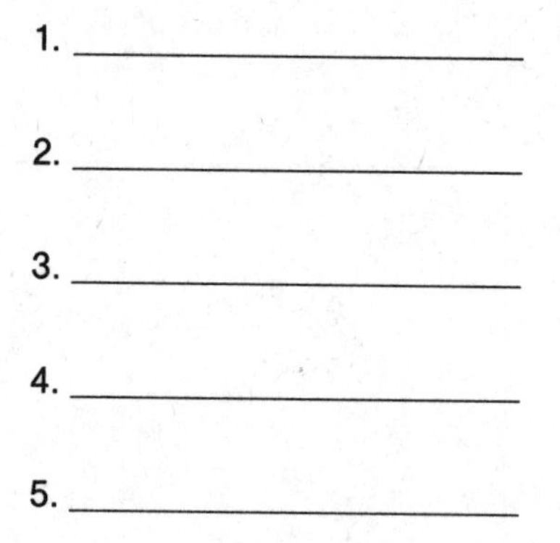

> **You can rewrite the story in your nervous system by fluently speaking the secret language of the body.**

Role Modelling

One of the most vital roles of a parent is that of teacher and role model. Of course, the most important things that they teach and model to their children happens through attunement and providing for their development needs in a responsive and consistent fashion. But in addition to that, parents support and teach children in feeling safe to explore the world, try new things, learn new skills and see and interpret the things that are happening in their lives in insightful and helpful ways. Those lessons can be as simple as training a helpful behaviour like brushing their teeth before bed or teaching them to say please and thank you. But it can also be more nuanced. For example, when a four-year-old girl gets upset because she wants a toy but another child is playing with it, the attunement component is comforting the child's upset and redirecting them to another toy to soothe them, but the teaching is in the gentle explanation to the child that says: 'I know it's hard to not be able to play with the toy you want. But we can practise being patient and taking turns, so everyone gets a chance to enjoy it.' And in order for these useful behaviours and life lessons to be transmitted, they have to be modelled many times over years (if this sounds like how neuroplasticity works, it's because it is, as all learning follows the laws of neuroplasticity).

In bonding, taking on the responsibility of a role model can be a very powerful way of training and reminding your nervous system to deploy more useful perspectives and behaviours in the place of old coping mechanisms. For example, every time Elsa noticed herself getting triggered she would summon her attuned self to connect with her inner child and model the following for her: 'I know that when we were little, Mom didn't know how to give us what we needed. But we aren't little anymore. We know that we are a strong, fierce and confident adult. Mom was not trying to hurt us, but she did and together we are repairing the hurt and healing the wound. I am here for you. Together, we can do this. You've been doing such a good job, I am proud of you and I'm here for you.' The more you can be a helpful and caring role model for your inner child, the more you will be providing new and useful modelling, establishing responses and behaviours that will improve your life.

These techniques, embracing 'now is not then', resourcing attachment, updating your story and taking on the role of parent-teacher, are powerful in and of themselves. 'Now is not then' rescues your inner child from the past where she or he has been stuck. Resourcing attachment lets your inner child deeply feel the attention, security, belonging and validation, connection and responsiveness that it so desperately needs. Updating your story helps to counteract the longstanding narratives and limiting beliefs that cut you off from the love and connection that, whether small or big, is actually around you. And being a parent-teacher to your inner child provides them with the guidance, modelling and support that they need to learn new and better ways of being. Together, they are a powerful mix of embodied and cognitive co-regulation tools that have an impact that are greater than the sum of their parts and hold the key to transforming your perception of self, others and the world.

Anchoring the Bond

Think about a song that has a particular meaning for you. It likely has this meaning because it accompanied you during a particular memorable time in your life. Maybe it kept you and your best friends company on that amazing adventure you had backpacking through Europe all those years ago, or maybe it helped keep your spirits high during those difficult times after you lost your job. Or maybe it's the song that, one winter, kept you company every day on your drive to drop your kids off at school. And maybe the first time you heard it, you loved it so much that you replayed it many times because of how it made you feel. As time passed, and you repeatedly heard that song, it acquired this power to take you back to that particular time. So, now, whenever you encounter the song, even in unrelated contexts, it acts as a cue for that same emotional response, evoking the same state in your body related to the first few times you listened to it. This innate ability of our brains to unconsciously establish connections between stimuli and our emotional states is called *anchoring*.

Anchoring is used in NLP and is a tool that aims to establish a robust and deliberate connection between a specific sensory stimulus, often a touch, word, movement or visual cue, and a corresponding emotional or nervous system state. By repetitively pairing these elements, you can effectively construct neural pathways in the brain, hardwiring the association between the stimulus and the new desired response. In fact, anchoring takes advantage of the same neurological processes that create a trigger, except in this instance those processes are used to 'trigger' (anchor) a positive state and adaptive responses to a cue instead of negative states and maladaptive responses. Psychotherapist Deb Dana calls these positive triggers 'glimmers'. In the same way a trigger can instantaneously dysregulate you, when an anchor is practised enough and the association between the cue and positive state you are choosing becomes strong, the anchor can *rapidly* help you shift states and become a short-cut to regulation.

With bonding you will use anchoring to both strengthen and accelerate the bond between your attuned adult self and your developmental self to redesign the neural networks that have kept you stuck in survival mode. This highly intentional healing and repair will help you shift the emotional and behavioural reactions of your inner child and developmental nervous system and help you finally experience what it is like to feel whole.

Let's learn to bond.

Let's take a moment to summarise the knowledge on Bonding with your inner child:

- Children develop internal working models (IWMs) about whether they can trust people based on whether they are treated in an attuned or mis-attuned way by their primary caregivers.
- When our IWM is organised around unworthiness, mistrust, fear of others and that the world is a dangerous place, our nervous system is incapable of truly feeling safe and can become chronically dysregulated.
- It's helpful to rely on your positive relationships to contradict the negative aspects of your IWM.
- Bonding is about being an attuned adult and re-shaping your IWM from one that is governed by worthlessness and fear to one of profound worthiness and love.

Practices 9

Redesign: Bonding

Learn to re-pattern your internal working model and mend your attachment injuries through consistent and responsive bonds of love

In these practices you will learn to embody thriving safety and heal the nervous system of your inner child through polyvagal-based co-regulation exercises, resourcing and NLP techniques.

Why Do It?

To heal the parts of you that you thought were lost, broken and buried.

When to Do It

Whenever you feel triggered, remember that you speak the secret language of your body. Use it!

Tips before you begin

- Responsiveness and consistency are key. The more you attend to your developmental needs, the more regulated your nervous system will become.
- Attention is nourishment for children. Think of your bonding practice as quality time with your inner child.
- Love yourself regularly.

As with all the Redesign practices in A I R, bonding is about using the principles of neuroplasticity to create lasting changes in your brain and nervous system. We do this by practising tending consistently until a regulated developmental self becomes your new default. Therefore bonding is consistent and responsive tending. We invite you to make tending a regular practice in your life and amplify it by incorporating the bonding techniques discussed in this chapter as well as anchoring (below).

Practice 1: Anchoring

Note that before you do this exercise, you will need to have done Practice 2 from the previous chapter.

Find and Label the Emotion

1. Find a quiet and comfortable place to sit in.
2. Notice what feelings are present and how these are showing up in BASE-H.

3. Tune into your inner child, and notice what part of your body they immediately show up in. For example, you may feel a tight throat, an empty sensation in your belly.
4. Label the emotion and the location of it.

Identify the Need

1. Ask your inner child what they need. For example, since you may already know this from the previous practice section, you may simply hear 'I need attention'.
2. Through all the practices you have learned thus far, give your inner child what they need through the power of your social engagement system. For example, they may need attention, and you give them a hug, make eye contact, tell them you love them and that they are doing a great job.
3. Notice how your inner child feels now.

Anchor the Bond

1. You are now going to anchor this bond into your body by associating it with a cue. Choose what cue you want to provide it with.
    a. This can be an NSM or hugging yourself, pressing your feet into the ground, putting your palms over your eyes or swaying.
    b. The cue you choose should feel good in your body and be easy to do and remember.
2. Once again, tune into how your inner child feels when you meet its needs with your social engagement system.
3. Activate the cue for the bond by performing an action inspired by the suggestions above in 1a.

4. Feel that immediate cascade of positive feelings, sensations and emotions.
5. Practise cueing your bond up to three or four times, until the desired feeling is high in intensity.
6. Check in with the original trigger that activated your inner child and notice how much less loud it is now that you are an attuned adult.

Now that the bond is anchored in the body, when your inner child is triggered, instead of always doing a tending exercise, you will use the cue you have anchored and instantly activate connection and provide regulation to your inner child. You will literally be re-designing your neurology and repairing your unhealed wounds.

Practice 2: Bonding

1. Find a quiet and comfortable place to sit in.
2. Connect with your inner child by:
    a. Cueing your anchor.
    b. Or engaging in the *providing for your inner child practice* from Chapter 8.
3. Resource attachment.
    a. Imagine your inner child surrounded by those that represent positive, loving and supportive relationships in your life for 2–3 minutes.
    b. As an attuned adult, give your inner child a warm hug.
    c. Notice and label how this feels in your body.
4. Now is not then.
    a. Invite your inner child to notice how it feels to know that 'now isn't then', and that now our needs are being met.
    b. Invite them to be 'in the now', with your attuned adult and positive relationships and to feel that it 'belongs'.
    c. Notice and label how this feels in your body.
5. Update your story.
    a. Validate the default ways your developmental self has adapted to help keep you safe.
    E.g. *'I know that when Gillian [our wife] forgot to get the groceries we asked for, it reminded you what it felt like when our mom neglected us.'*
    b. Through dialogue, have your Observing Self share with your inner child the updated story as well as some of the diamonds in your life for 2–3 minutes.
    E.g. *'But the truth is that our wife is not our mom, and Gillian tends to our needs all the time. Remember just*

yesterday when she surprised us by bringing lunch to us at work? Didn't that feel so nice? We are so so cared for. It's OK if she makes a mistake now and then.'

c. Notice and label how this feels in your body.

6. Role model.
   a. Remind your inner child that you now possess all the qualities of consistent and responsive attuned attachment.
   b. Remind your inner child about the new more helpful ways of perceiving and responding to triggers that you have been practising.
   c. Visualise yourself responding together in this new way.
   d. Notice and label how this feels in your body.

Experiment and make this practice your own: There is no 'right' way to do this work. Use your imagination, try different things (even if they feel silly) and combine the above ingredients in a way that deeply resonates for you.

Make Bonding a Part of Your Life

The goal is to redesign your internal working model by making caring for and parenting your inner child a way of life. To do this, we invite you to:

1. Stay attuned to your needs.
2. Respond to them in real time with your cue.
3. Anytime an old pattern comes up and is resolved, remember to be the attuned adult your inner child needs, so you learn how to be the adult you've always wanted to be.

Notes:

....

....

....

....

....

....

....

....

Conclusion

Living the Language of the Body

After they had explored all the suns in the universe, and all the planets of all the suns, they realized that there was no other life in the universe, and that they were alone. And they were very happy, because then they knew it was up to them to become all the things they had imagined they would find.

Lanford Wilson, *Fifth of July*

Einstein once said, 'everything should be made as simple as possible, but no simpler'. In this book, we have endeavoured to do just that. To take the vast complexity of the human experience as seen through the lens of the nervous system and reduce it to its most fundamental elements so that you can understand it, work with it and heal yourself. Simplicity is always at the heart of complexity. In the same way that Sir Isaac Newton's three laws of motion explains the movement of everything from birds in flight to the spinning of galaxies, A I R applied to your mind, body and human realms can explain and change your entire nervous system and, by extension, transform your life.

Living the language of the body is intentionally committing to maintain awareness of your mind, body and humanness and use A I R whenever your nervous system is communicating to you (via a symptom) that something needs your attention. As described in the Reader's Guide at the beginning of this book, everything you have learned is applied in practice like this:

1. I notice that I am dysregulated because of a symptom (e.g. negative self-talk, anxiety, physical tension, unexplained fatigue).
2. I use an **awareness** technique like *listening* or *translating* to tune into the language of my body to get a deeper understanding of what I am experiencing and what triggered my dysregulation.
3. I use an **interruption** technique like *switching* or *modifying* to disrupt and soothe my dysregulation.
4. I use a **redesign** technique like *distancing* or *bonding* to stimulate neuroplasticity and reinforce the new regulated response to the trigger.

Of course the art of this science is to learn - through experimentation and the ups and downs of life - what exactly your body is communicating to you and what combination of A I R techniques is required to meet your own needs. There is no one size fits all prescription or protocol because even though our lives are governed by the same nervous system laws, they present in each of us in very different ways. But as human beings, we have very common experiences and therefore we can provide you with some examples of how this work is done in real life to map onto your own life, to reiterate the best practices to help you regulate your nervous system.

Some Examples of Real Life Nervous System Regulation

Jane's Back Pain

Jane notices that her chronic low back pain is acting up at work. She knows that is a message from her body and takes a moment to stop what she is doing and start *listening* (Chapter 1, Practice 1). As she tunes into BASE-M she becomes aware that along with the back pain she is feeling a lot of pressure and fear. She does the *modifying practice of downregulating the sympathetic* (Chapter 5, Practice 2) which makes her feel about 50 per cent better. Jane then *steps outside herself* (Chapter 3, Practice 1) and her Observing Self asks her inner child why she is feeling so much pressure and fear. Her inner child tells her that she feels like all the responsibilities she has are overwhelming and that if she fails to please everyone she'll be abandoned, worthless and unloved. Jane then *provides for her inner child* (Chapter 8, Practice 2) and her pain disappears and she feels that she is nearly 100 per cent regulated and capable of going back to work. Before she does, she takes 2 minutes to do the *bonding practice of updating her story* (Chapter 9, Practice 2) to reinforce for her nervous system that she is a competent adult and not a child anymore. Because Jane is familiar with the exercises, has practised them and knows which ones help her instantly shift state, all in all this takes her no longer than 5 minutes.

Bob's Depression

Bob is supposed to be getting ready to go to his friend's birthday party but notices that he can't find the motivation to get off the couch and recognises that he is feeling depressed. From practising the *translating* techniques of *knowing your default response* (Chapter 4, Practice 2) he is able to tune into that fact that he is stuck in his default pathway of dorsal shutdown. He then *upregulates the dorsal* (Chapter 5, Practice 3) and finds that he has a lot more energy and is less stuck (if he is short on time, he stops here and gets ready to go to the party). With his new-found energy he takes a moment to do *attuning to your developmental need* (Chapter 7, Practice 2) and discovers that his developmental nervous system is remembering all the times he wasn't invited to parties as a child and the hurt he felt during those times. He then takes a moment to bond with this inner child and do the *now is not then and resourcing attachment* techniques (Chapter 9, Practice 2) to help his inner child feel the love and wonderful social life he has now. Bob has practised moving out of depression for a few months now, so it doesn't feel as difficult as the first time. His nervous system is learning to keep him safe in new ways like feeling whole and connected to others. He may or may not feel 100 per cent regulated before the party, but the exercises get him to a place of mobilisation which helps him get to his friend's birthday and feel even more regulated through the course of the night.

Lucy's Anxiety

Today is the day of Lucy's family reunion. She barely slept the night before and upon waking she notices a significant spike in the feeling of anxiety. Drawing upon the *translating* exercise of *learning the body's responses* (Chapter 4, Practice 1), she recognises she's falling into the fight or flight state and anxiety response. Using the *switching* technique of *push* (Chapter 3, Practice 1) followed by the *modifying* technique of *regulating to ventral* (Chapter 4, Practice 1), Lucy starts to shift her anxiety, noticing a significant reduction in the feeling of tension, nervousness and fear. With her anxiety now at a more manageable level in the experience of her mind and body, Lucy uses the *settling* technique of *soothe your body* (Chapter 6, Practice 2). She realises that her current anxiety is partly linked to childhood events where she felt overlooked, misunderstood and unheard. To continue working on the repair of these past experiences, she uses the *tending* exercise of *providing for your inner child* (Chapter 8, Practice 2) to help reassure her younger hurt self, making her feel cared for and loved. Lucy now feels more confident in setting boundaries when she needs to and in remembering that family members will say and do things often by default (because they have their own nervous systems and coping mechanisms) and feels more confident in not taking remarks or comments personally. As the time for the gathering approaches, Lucy, armed with a newfound sense of repair and inner strength, is able to join her family with ease. Her nervous system is adapting, learning to find safety and connection in these once-challenging social scenarios. While anxiety may come back to her as a whisper instead of a scream, the techniques she's practised have brought her to a place of change and redesign, enabling her to be flexible and regulated.

Best Practices Reminder

1. The laws of neuroplasticity state that the more consistently you do something the faster your brain will change and learn the new skills. For this reason, we encourage you to practise A I R techniques as often as possible and integrate them into your daily rhythm so that it doesn't feel like a task, but a healing way of life. We always say that it's not about making space for the exercises, but about moving through life in a new way that encompasses the healing tools in your everyday experience.

2. And with that in mind our invitation is not to overdo it – pressure and perfectionism tend to inhibit learning and neuroplasticity. Recognising the importance of this work for your healing and being called to practise will always work better than feeling that it's urgent and pushing yourself to do it. In fact as you do this work, you will find that perfectionism and pressure were messages coming from your body in the language of your nervous system and when you understand the messages you will be able to shift those patterns within you.
3. The best time to practise is *in response* to a symptom like anxiety, stress, self-criticism, self-doubt, pain, procrastination or depression.

Tips

1. The techniques in this book were taught in a particular order for *learning* purposes but they do not need to be *applied* in that same order when you use them in your life. In practice, depending on the trigger you are experiencing, you may find it more effective to use *attuning to your developmental needs* (Chapter 7, Practice 2) followed by *regulate to ventral* (Chapter 5, Practice 1) and completing with *anchoring* (Chapter 9, Practice 1). Explore, experiment, learn to trust your intuition and do what works for you.
2. One of our teachers says that symptoms are sacred. Every symptom is there for a reason and has a message to convey. They are always invitations and opportunities to expand your awareness and deepen your understanding of yourself and capacity to provide what you need. Embrace

them. Learn from them. Care for them. Ultimately your symptoms are your greatest teacher and will set you free.

3. If you find yourself very activated in sympathetic and shutdown in dorsal, use the interruption techniques of *switching* from Chapter 2 or *modifying* from Chapter 5 to move you into a more regulated state before trying the redesign techniques. Feeling stuck or unable to regulate, especially in the first instances of doing this work, is completely normal. Just like it may take some time for you to trust a stranger, your nervous system may need time to trust this new version of you.
4. Although it is very helpful to use an awareness, interruption and/or redesign technique in response to dysregulation, this may not always be practical. It's perfectly OK for you to apply any single practice you like to help regulate in the moment. For example, doing the *five senses experience* (Chapter 2, Practice 5) or *soothe your body* (Chapter 6, Practice 1) or any of the *switching* and *modifying* techniques as standalone techniques can be great ways of regulating yourself when you are short on time.

And lastly, the most important tip of all - although this work is to help you regulate your nervous system and heal yourself - is to know yourself, love yourself, feel whole within yourself, say no when you want to say no, do the things you want to do, show up as who you want to show up as, choose the things in life that bring you joy, have the courage to choose yourself every day, and feel comfortable and at home in your own body. When you feel all these things, and are able to live them through living the language of your body, tending to your unmet needs and repairing deep wounds, you rewrite the story of your mind and body and heal. You know yourself. You know how to shift these experiences. You know how to speak to your nervous system and whisper, in its language, of what it needs to feel safe and be healed.

A Note from Jen

The day I began healing was the day I realised that I didn't have to fit into a story that wasn't mine anymore, that I didn't have to make myself smaller, kinder, nicer, better for other people in order to feel worthy. That I didn't have to abandon myself in order to feel seen. That I didn't have to achieve anything to feel 'enough'. The day I began healing was the day I realised I could stop apologising for being me and that the uncomfortable consequences of not being perfect would never be greater than the feeling of being free. Choosing me, over what I thought you needed me to be in order to love me, healed me. Coming home to myself and feeling the permission of leaning into doing things my way healed me. Enoughness, worthiness, self-compassion, self-love – did not come easy, but a seed was planted and they grew inside me over time replacing the hollow darkness, hurt and anger that was once there. Although doctors told me that my chronic illness was 'in my head' and that I would never fully recover, I always knew how real it felt to me and I always knew that I would find my way out.

To you, dear reader, for whom this manual of healing was created and who may be wrestling with your own struggles, I offer this reflection: I see you in your entirety, in the hurt that you carry and the resilience that you embody. Your experiences are real, they matter and, yes, they are not easy. But amidst the seeming impossibility, please remember that healing, though challenging, is within reach. It's a journey that I know you're already familiar with, of returning to you, of finding strength in vulnerability and of embracing your authentic truth, no matter how daunting it feels.

Your path to healing is uniquely yours, and it's marked by the courage to face each day, to nurture the seed of hope within you and to slowly but surely reclaim your wholeness. Remember, in this journey of healing, you are not alone. Whatever you're experiencing, dear reader, I see you. It's real and it's marvellous because it's what makes you so special, so strong and so aware. Those of us who are required to dive deep and do the work are the ones who will emerge with greater wisdom than any book, course or university can ever teach. Healing is such a unique fingerprint - what works for one may not work for another - but I want you to embody - in your bones - that even when it feels impossible, healing is possible.

As you read these final pages, I want to extend my deepest gratitude to you for journeying with us through the chapters of this book, your healing manual. On that note, I invite you to take a moment and do one more practice with me.

1. Pause.
2. Take a moment to recognise just how brave and resilient you are.
3. Take a deep breath in, hold your breath at the top for 6 seconds.
4. Exhale slowly for as long as you can emptying your lungs.
5. Give yourself a hug and feel hugged in this warm embrace.
6. Stay here as long as you need as I share words with you. I know the following to be true: **you are more resilient than you know, more strong than your toughest days, more loved than you feel and more capable than you believe.**

Your body holds stories, but through this work you get to be the author who decides how they are told. Be gentle with yourself, for you are not surviving anymore; you are deep-diving, exploring, transforming, and in that exploration and transformation of the raw and fierce complex beauty of being (alive) you will learn to move through life unapologetically as YOU. I'm proud of you. If we ever meet in person please stop and say hi so I can give you a big hug.

With heartfelt compassion and a deep admiration of you and your truth,

Jen
Bali, Indonesia
20 November 2023

A Note from Karden

As a child, I took refuge from the world in fantasy and sci-fi stories. Each time I'd watch *Star Wars* I wished I was a Jedi and would try to use the Force to summon the TV remote into my hand (which tragically never worked). I would read comic books and daydream of being a hero and having superpowers. In my teens, I read Frank Herbert's *Dune* and was fascinated by the way the Bene Gesserit had such fine-tuned control of their bodies that they could literally tell their cells what hormones to make. In my twenties, I began practising meditation, tai chi, yoga and other non-Western practices, and my motivation was in no small part derived from wanting to be a real-life Bene Gesserit superhero Jedi. Although I still haven't figured out how to use the Force to make the remote control fly into my hand, nor have I learned to fly or shoot spider webs, learning to listen and speak the language of my body has given me the closest thing to superpowers one can have in real life. After all, when both conventional and alternative medicine were unable to help me with my agonising back pain but then, by discovering that it was my nervous system (and not my vertebral discs) and that feeling my repressed emotions made the pain vanish, if that isn't a superpower, I don't know what is.

When people are stuck and powerless, they will frequently use the phrase 'this is just the way I am'. When I was stuck, I said it all the time, almost like a badge of honour. In fact, much of my personality organised itself around justifying itself to myself and others. What I have learned is that 'this is just the way I am' is a phrase that kills all possibility of learning, change and growth. It's a statement that creates neuro-concrete instead of neuroplasticity. For half of my life, I was a tangled knot of repressed attachment wounds, chronic pain and coping mechanisms that made me a deeply insecure, sad, angry, unfeeling and unkind, well, jerk. Through a combination of desperation, inspiration and pure blind luck, I stumbled into the secret language of my body and nervous system healing. I have used these superpowers for far more than just resolving my pain. I have now spent the second half of my life as a man, friend, husband and father who feels. Feels for himself and feels for others. I live a life of consciousness and intention instead of repression and reactivity. I rarely have back pain and when I do I receive it as a sacred symptom that a part of me needs attention and support. My life experience along with the thousands of others that Jen and I have helped transform with this work has proven to me that there is no such thing as 'this is just the way I am', and I invite you to consider that there is, in fact, no limit to your ability to heal and become the person you want to be.

We start this work to get out of some form of pain, but we stay with this work to transcend. We heal the injuries and coping mechanisms that cause our lives to be an act of survival so we can transcend to one where we are actually thriving. We liberate ourselves from the shallow world of our mind and thoughts so we can transcend into the richness and depth of living a fully embodied life. We become the one to break the cycle of intergenerational trauma in our family so we can transcend our past and provide a different foundation for our children. We may start this work to heal, but we stay with this work because it allows us to extract ever more vibrancy out of this wild and beautiful life we get to live.

You are now awakened to the language of your body which once awakened, rarely goes back to sleep. There is no beginning or end to this journey. There is no steady state. It is always ebbing and flowing and will continue to do so for the rest of your life. This awakening brings with it the amazing opportunity to cultivate your own superpowers for healing yourself and transforming your life. But as Spiderman's uncle said, 'with great power, comes great responsibility'. In this case, that responsibility is to honour the messages of your nervous system and lead a life that is in alignment with your true self. It is my dearest wish that you choose to do so.

Karden
West Stockbridge, Massachusetts
20 November 2023

For more support on healing your nervous system and to connect with our global community, check out:

www.somiainternational.com/book

On Instagram
@somiainternational
@iamjenmann
@kardenrabin

Notes

Introduction

1. ZERO TO THREE, When does the fetus's brain begin to work? (www.zerotothree.org/resource/when-does-the-fetuss-brain-begin-to-work/)
2. H. Klumpp, M.K. Keutmann, D.A. Fitzgerald, S.A. Shankman and K.L. Phan, Resting state amygdala-prefrontal connectivity predicts symptom change after cognitive behavioral therapy in generalized social anxiety disorder. *Biology of Mood and Anxiety Disorders*, 4, 1 (2014): 14.
3. B.S. McEwen, Brain on stress: How the social environment gets under the skin. *Proceedings of the National Academy of Sciences*, 109, supplement_2 (2012): 17180–17185.
4. J. Guidi, M. Lucente, N. Sonino and G.A. Fava, Allostatic load and its impact on health: a systematic review. *Psychotherapy and Psychosomatics*, 90, 1 (2021): 11–27.
5. M.B. Yunus, Role of central sensitization in symptoms beyond muscle pain, and the evaluation of a patient with widespread pain. *Best Practice & Research: Clinical Rheumatology*, 21, 3 (2007): 481–497.
6. P. Mateos-Aparicio and A. Rodríguez-Moreno, The impact of studying brain plasticity. *Frontiers in Cellular Neuroscience*, 13 (2019): 66.
7. Á. Carlos-Reyes, J.S. López-González, M. Meneses-Flores, D. Gallardo-Rincón, E. Ruíz-García, L.A. Marchat, H. Astudillo-de la Vega, O.N. Hernández de la Cruz and C. López-Camarillo, Dietary compounds as epigenetic modulating agents in cancer. *Frontiers in Genetics*, 10 (2019).
8. D.C. Lin, Exercise impacts the epigenome of cancer. *Prostate Cancer Prostatic Disease*, 25 (2022): 379–380.

Part I – Mind

Chapter 1: Awareness – Listening

1. Cleveland Clinic, Cerebral cortex (https://my.clevelandclinic.org/health/articles/23073-cerebral-cortex).
2. University of Queensland, The limbic system (https://qbi.uq.edu.au/brain/brain-anatomy/limbic-system).
3. A.J. Crum, W.R. Corbin, K.D. Brownell and P. Salovey, Mind over milkshakes: mindsets, not just nutrients, determine ghrelin response. *Health Psychology*, 30, 4 (2011): 424–429; discussion 430–431.
4. J. Marchant, *Cure: A Journey into the Science of Mind over Body*. Crown Publishing, 2016.

Chapter 2: Interruption – Switching

1. N.P. Gothe, I. Khan, J. Hayes, E. Erlenbach and J.S. Damoiseaux, Yoga effects on brain health: a systematic review of the current literature. *Brain Plasticity*, 5, 1 (2019): 105–122 (www.ncbi.nlm.nih.gov/pmc/articles/PMC6971819/).
2. D. Keltner, Hands on research: the science of touch. *Greater Good Magazine*, 2010 (https://greatergood.berkeley.edu/article/item/hands_on_research).
3. D. Juhan, *Job's Body*. Station Hill Press, 2003.
4. T. Shafir, Using movement to regulate emotion: neurophysiological findings and their application in psychotherapy. *Frontiers in Psychology*, 23, 7 (2016): 1451.
5. Shafir, Using movement to regulate emotion, 1451.

Chapter 3: Redesign – Distancing

1. S. Freud, *New Introductory Lectures on Psycho-Analysis. The Standard Edition of the Complete Psychological Works of Sigmund Freud, Volume XXII (1932–1936): New Introductory Lectures on Psycho-Analysis and Other Works*, 1933. 1–182.
2. J. Guttman, How observing your ego can improve your social functionality. *Psychology Today*, 2021 (www.psychologytoday.com/us/blog/sustainable-life-satisfaction/202107/how-observing-your-ego-can-improve-your-social).
3. Bessel van der Kolk, *The Body Keeps the Score: Brain, Mind, and Body in the Healing of Trauma*. Penguin, 2015.

4. M. Koenigs and J. Grafman, Posttraumatic stress disorder: the role of medial prefrontal cortex and amygdala. *Neuroscientist*, 15, 5 (2009): 540–548.
5. A. Etkin and T.D. Wager, Functional neuroimaging of anxiety: a meta-analysis of emotional processing in PTSD, social anxiety disorder, and specific phobia. *American Journal of Psychiatry*, 164 (2007): 1476–1488.
6. Koenigs and Grafman, Posttraumatic stress disorder.
7. L.A. Kilpatrick, B.Y. Suyenobu, S.R. Smith, J.A. Bueller, T. Goodman, J.D. Creswell, K. Tillisch, E.A. Mayer and B.D. Naliboff, Impact of mindfulness-based stress reduction training on intrinsic brain connectivity. *NeuroImage*, 1, 56 (2011): 290–298.
8. Van der Kolk, *The Body Keeps the Score.*
9. P.B. Sharp, B.P. Sutton, E.J. Paul et al., Mindfulness training induces structural connectome changes in insula networks. *Scientific Reports*, 8 (2018): 7929.
10. Walpola Rahula, *What the Buddha Taught*. Grove Press, 1974.
11. S. Cope, *Yoga and the Quest for the True Self*. Bantam, 2000, p. 186.
12. ScienceDaily, How curiosity changes the brain to enhance learning (www.sciencedaily.com/releases/2014/10/141002123631.htm).

Part II – Body

Chapter 4: Awareness – Translating

1. D. Mobbs, C.C. Hagan, T. Dalgleish, B. Silston and C. Prévost, The ecology of human fear: survival optimization and the nervous system. *Frontiers in Neuroscience*, 9 (2015).

Chapter 5: Interruption – Modifying

1. C. Duhigg, *The Power of Habit*. Random House, 2014.
2. University of Queensland, How Your Brain Makes and Uses Energy (https://qbi.uq.edu.au/brain/nature-discovery/how-your-brain-makes-and-uses-energy).
3. Bessel van der Kolk, *The Body Keeps the Score: Brain, Mind, and Body in the Healing of Trauma*. Penguin, 2015, p. 270.
4. Centre for Neuro Skills, Ten principles of neuroplasticity (www.neuroskills.com/brain-injury/neuroplasticity/ten-principles-of-neuroplasticity/).
5. V.A. Pavlov and K.J. Tracey, The vagus nerve and the inflammatory reflex - linking immunity and metabolism. *Nature Reviews Endocrinology*, 8, 12 (2012): 743–754.

6. S. Breit, A. Kupferberg, G. Rogler and G. Hasler, Vagus nerve as modulator of the brain–gut axis in psychiatric and inflammatory disorders. *Frontiers in Psychiatry*, 13, 9 (2018): 44.
7. F. Cerritelli, M.G. Frasch, M.C. Antonelli, C. Viglione, S. Vecchi, M. Chiera and A. Manzotti, A review on the vagus nerve and autonomic nervous system during fetal development: searching for critical windows. *Frontiers in Neuroscience*, 15 (2021).
8. C. Darwin, *The Expression of the Emotions in Man and Animals*. 2012. p. 74.

Chapter 6: Redesign – Settling

1. A. Vaish, T. Grossmann and A. Woodward, Not all emotions are created equal: the negativity bias in social-emotional development. *Psychological Bulletin*, 134 (2008): 383–403.
2. S. Soroka, P. Fournier and L. Nir, Cross-national evidence of a negativity bias in psychophysiological reactions to news. *Proceedings of the National Academy of Sciences*, 116, 38 (2019): 18888–18892.
3. P. Levine, *In an Unspoken Voice*. North Atlantic Books, 2010, p. 125.
4. L. Artigas and I. Jarero, The butterfly hug method for bilateral simulation, 2014 (https://emdrfoundation.org/toolkit/butterfly-hug.pdf; https://www.counselingconnectionsnm.com/blog/try-the-butterfly-hug-to-help-with-ptsd-symptoms).
5. National Institute for the Clinical Application of Behavioral Medicine (NICABM), Two simple techniques that can help trauma patients feel safe (www.nicabm.com/trauma-two-simple-techniques-that-can-help-trauma-patients-feel-safe/comment-page-3/).

Part III – Human

Chapter 7: Awareness – Attuning

1. L. Jeličić, A. Veselinović, M. Ćirović, V. Jakovljević, S. Raičević and M. Subotić, Maternal distress during pregnancy and the postpartum period: underlying mechanisms and child's developmental outcomes – a narrative review. *International Journal of Molecular Sciences*, 23 (2022): 13932; E. Merlot, D. Couret and W. Otten, Prenatal stress, fetal imprinting and immunity. *Brain, Behavior, and Immunity*, 22, 1 (2008): 42–51.
2. A.M. Graham, J.H. Pfeifer, P.A. Fisher, S. Carpenter and D.A. Fair, Early life stress is associated with default system integrity and emotionality

during infancy. *Journal of Child Psychology and Psychiatry*, 56 (2015): 1212–1222.

3. O. Wlodarczyk, M. Schwarze, H.J. Rumpf, F. Metzner and S. Pawils, Protective mental health factors in children of parents with alcohol and drug use disorders: a systematic review. *PLoS One*, 12, 6 (2017): e0179140.
4. Centers for Disease Control and Prevention (CDC), About the CDC-Kaiser ACE Study (www.cdc.gov/violenceprevention/aces/about.html).
5. Bessel van der Kolk, *The Body Keeps the Score: Brain, Mind, and Body in the Healing of Trauma*. Penguin, 2015, p. 148.
6. Z. Hu, A.C. Kaminga, J. Yang, J. Liu and H. Xu, Adverse childhood experiences and risk of cancer during adulthood: A systematic review and meta-analysis. *Child Abuse & Neglect*, 117 (July 2021): 105088.
7. L. Mandelli, C. Petrelli and A. Serretti, The role of specific early trauma in adult depression: A meta-analysis of published literature. Childhood trauma and adult depression. *European Psychiatry*, 30, 6 (September 2015): 665–680.
8. S.R. Dube, D. Fairweather, W.S. Pearson, V.J. Felitti, R.F. Anda and J.B. Croft, Cumulative childhood stress and autoimmune diseases in adults. *Psychosomatic Medicine*, 71, 2 (2009): 243–250.
9. V. Rattaz, N. Puglisi, H. Tissot and N. Favez, Associations between parent-infant interactions, cortisol and vagal regulation in infants, and socioemotional outcomes: a systematic review. *Infant Behavior and Development*, 67 (2022): 101687.
10. K. Cherry, Harry Harlow and the nature of affection, Verywellmind, 2023 (www.verywellmind.com/harry-harlow-and-the-nature-of-love-2795255).
11. D. Juhan, *Job's Body*. Station Hill Press, 2003, p. 208.
12. C.A. Obeldobel, L.E. Brumariu and K.A. Kerns, Parent–child attachment and dynamic emotion regulation: a systematic review. *Emotion Review*, 15, 1 (2023): 28–44.
13. Cleveland Clinic, The 4 attachment styles and how they impact you, 2022 (https://health.clevelandclinic.org/attachment-theory-and-attachment-styles/).

Chapter 8: Interruption – Tending

1. A. Rosengren, S Hawken, S. Ounpuu, S. Kliwa, M. Zubaid, W.A. Almahmeed, K.N. Blackett, C. Sitthi-Amorn, H. Sato, S. Yusuf, INTERHEART investigators. Association of psychosocial risk factors with risk of acute myocardial infarction in 11119 cases and 13648 controls from 52 countries (the INTERHEART study): case-control study. *Lancet*, 364, 9438 (2004): 953–962.
2. G. Noppe, E. van den Akker, Y. de Rijke et al., Long-term glucocorticoid concentrations as a risk factor for childhood obesity and adverse body-fat distribution. *International Journal of Obesity*, 40 (2016): 1503–1509.
3. K. Shankardass, M. Jerrett, J. Milam, J. Richardson, K. Berhane and R. McConnell, Social environment and asthma: associations with crime and No Child Left Behind programmes. *Journal of Epidemiology and Community Health*, 65, 10 (2011): 859–865.
4. S.R. Dube, D. Fairweather, W.S. Pearson, V.J. Felitti, R.F. Anda and J.B. Croft. Cumulative childhood stress and autoimmune diseases in adults. *Psychosomatic Medicine*, 71, 2 (2009): 243–250.
5. A. Mehrabian, Nonverbal communication. *Nebraska Symposium on Motivation*, 19 (1971): 107–161.
6. L. Castañón, Global collaboration led by Stanford researcher shows that a posed smile can improve your mood. *Stanford News*, 2022 (https://news.stanford.edu/2022/10/20/posing-smiles-can-brighten-mood/).

Chapter 9: Redesign – Bonding

1. R. Hanson and R. Mendius, *Buddha's Brain: The Practical Neuroscience of Happiness, Love & Wisdom*. New Harbinger Publications, 2009, p. 7.

About the Authors

Jennifer Mann, a mind-body practitioner and expert in the field of treating psychophysiological disorders, is the co-founder of Somia and the Heal Program. While at the pinnacle of her career in holistic health and wellness, without warning, overnight she became bedridden. She was diagnosed with multiple chronic illnesses, anxiety, pain and myriad 'unexplained' symptoms, and despite her training in neurorehabilitation and physiotherapy and study of biomedical sciences, Jen was at a loss to understand her illness, and conventional medicine offered no explanation. Told that life as she knew it was over, Jen refused to acknowledge that prognosis and became driven to find answers and reinvent her clinical practice.

In addition to achieving her professional qualifications, Jen's journey is defined by personal resilience, self-transformation and managing to do what the status quo said was impossible. With over a decade of studying the mind and body, Jen has emerged as a renowned clinician and an authority in integrating neurophysiology within holistic mind and body practices. Her work has been instrumental in helping thousands recover from stress, trauma and multiple chronic illnesses rooted in nervous system dysregulation.

Her personal recovery has profoundly informed her work; as a thought leader and guide to a worldwide online community, she brings her teaching, inspiration and empowerment to millions. She lives in Bali with her husband Yiannis and their precious son Leo.

Karden Rabin is the co-founder of Somia, co-creator of the Heal Program and a Somatic Experiencing practitioner who specializes in treating illnesses related to the nervous system. As an expert in the field of stress, trauma and psychophysiology, his work is dedicated to helping people recover from nervous system-related illnesses so they can live productive and fully expressed lives.

Having struggled with debilitating back pain for over a decade, and after traditional approaches hadn't helped, Karden managed to eliminate the pain by studying the psycho-emotional and neurological underpinning of chronic pain and stress-based disorders. Driven by his personal healing experience, he has helped thousands of clients all over the world manage and overcome their own pain, combining the principles of bodywork, brain retraining and somatic trauma therapies.

Over the past decade, Karden has developed and led programming for The Wounded Warrior Project; Starbucks; Kripalu Center for Yoga and Health; and is a regular contributor to Bessel Van Der Kolk's Trauma Research Foundation. Karden lives in the Berkshires of Massachusetts, with his lovely wife Gillian and their little women, Leia and Zelda.